WH

IN

AVIATION AND SPACEFLIGHT

From Abruzzo (Ben)
to
Zurakowski (Jan)

Compiled by

IAN WILSON

POCKET REFERENCE BOOKS

Published by:
Pocket Reference Books Publishing Ltd.
Premier House
Hinton Road
Bournemouth
Dorset BH1 2EF

First published 1996

Typesetting	Gary Tomlinson PrintRelate (Bournemouth, Dorset) (01202) 897659
Cover Design:	Van Renselar Bonney Design Associates West Wickham, Kent BR4 9QH
Illustrator:	John Warren
Printing and Binding:	RPM Reprographics Units 2-3 Spur Road Quarry Lane, Chichester West Sussex PO19 2PR Tel. 01243 787077 Fax. 01243 780012 Modem 01243 536482 E-Mail: rpm@argonet.co.uk

ISBN: 1 899437 51 7

Contents **Page**

Selected Charity — Inside Front Cover

Credits — 2

Contents — 3

Introduction — 4

Teasers — 5

Who's Who in Aviation and Spaceflight (From Abruzzo to Zurakowski) — 5 - 159

Glossary and Abbreviations — 160

History of Selected Charity — Inside Back Cover

ALL POCKET REFERENCE BOOKS INCLUDE A FREE FULL PAGE ADVERTISEMENT ON THE INSIDE FRONT COVER FOR A WELL-KNOWN CHARITY SELECTED BY THE AUTHOR

INTRODUCTION

Flying has become so commonplace that it seems hard to believe the aeroplane has been with us for less than a century.

Not even the most visionary of the early aviators would have predicted that within 75 years men, women and children would be travelling at twice the speed of sound, or that men would walk on the moon. Even more impressive than the advance in speed has been that of safety – leading airlines achieve around a million flights per fatal accident. For most passengers the most dangerous part of the journey is the drive to the airport. Yet this could not have been achieved without the extraordinary bravery, determination and skill of the men and women whose stories follow.

The common theme throughout these pages is that of courage. The pioneers learning to fly and to design at the same time, the wartime flyers, the test pilots and the trail-blazers as often as not rose to fame leaving a trail of wreckage behind them. As just one example, Blériot booked his place in history after building, and in most cases crashing, no less than ten other aeroplanes.

Inevitably there will be unfairness in the selection of names. No apology is made for the 'British bias' in the choice. Perhaps the greatest injustice is the omission of those who died seeking to make history or serve their countries and never reached the goals which would have won them renown.

The entries concentrate on what the people actually did. To make full use of the limited space, military ranks and decorations are included only where relevant, and in most instances principal forenames only are listed.

Ian Wilson

WHAT DO YOU KNOW ABOUT AVIATORS AND ASTRONAUTS?

TRY THESE TEASERS

Which one-eyed pilot flew round the world twice?

Which designer and pilot was unable to read or write?

Who was the first man on the moon?

Which pre-war pilot was offered a million pounds to bomb Hitler?

Who was the first person to exceed the speed of sound?

Whose unmanned balloon flew from Paris to Rome and caused international friction?

Who was the first to fly in America, after an earlier exploit in which he lost his trousers?

Who was the first to loop the loop?

Who claimed to have flown the Atlantic by mistake?

Who did fly a Lightning jet fighter by mistake?

Which two men flew 120 years before the Wrights?

Who built and flew, just once, a 700-seat flying-boat?

Whose 1935 airliner is still flying in some numbers?

Who built a fighter designed to ram enemy aircraft?

How did a British aerobatic pilot save himself after a wing failed?

Who made the first ripcord parachute descent?

Who built the first four-engine aircraft?

Why did popular hero Charles Lindbergh later become widely hated?

Who protested “I was hired to drive, not to fly?”

Which famous bomber pilot, and which jet engine pioneer, were originally rejected by the RAF?

Who took longer to fly across America than the record for walking the distance?

Which deaf Russian schoolteacher foresaw multi-stage manned spacerockets in 1903?

Who flew and sailed Gipsy Moths?

Which famous French company took its name from a Resistance code-word?

You will find the answers to these teasers in the following pages

ABRUZZO, BEN (1937-85) – first to cross the North Atlantic by balloon. At least 16 attempts had been made to cross the Atlantic by balloon, starting as far back as 1860.

Abruzzo and partner Maxie Anderson made their first bid to fly the Atlantic in 1977 in a helium balloon *Double Eagle*. Southerly winds carried them off course and they ditched, suffering from exposure, near Iceland.

With a third crew member, Larry Newman, they set out again on 12th August 1978 from Maine in *Double Eagle II*. They landed five days later in France, ending the first transatlantic crossing by a free balloon.

In 1981 he followed this triumph with a Pacific crossing by balloon.

Ironically he escaped the hazards of record-breaking balloon flights only to be killed, with his wife and four passengers, while flying a Cessna light aircraft.

ADER, CLÈMENT (1841-1919) – achieved a brief 'hop' in 1890. The engineer from Toulouse built his *Éole* in 1890. Following his studies of birds and bats, including flights of models, the full-size machine featured bat-shaped wings. It was powered by a 20 hp steam engine.

On 9th October 1890 he became airborne for about 165 ft (50 m). It was an achievement, but it does not qualify as sustained or controlled flight.

The French government commissioned him to build further aeroplanes, culminating in the *Avion III* with two steam engines. He tested it in 1897, but despite later extravagant claims, it never left the ground. Perhaps this was just as well, for he had provided no means of control - what did he plan to do if he became airborne? Ader was by no means alone in this respect. A surprising number of early aspiring aviators gave no thought to control or piloting technique.

Ader was handicapped by the weight of his steam engines and boilers. One legacy he left was to give the word 'avion' to the French language.

ALCOCK, SIR JOHN ('Jack') (1892-1919) – pilot of the first non-stop North Atlantic crossing by air. Born in Manchester, and colloquially referred to as 'The Manchester Lad', Alcock qualified as a pilot in 1912. He served in the RNAS in the First World War, being shot down and captured by the Turks in 1917.

In 1919 Vickers entered a Vimy bomber for a £10,000 prize offered by Lord Northcliffe, proprietor of the *Daily Mail,* for the first non-stop Atlantic flight. The Vimy was powered by two Rolls-Royce Eagle engines and modified to take extra fuel tanks. The crew were to be John Alcock and **Arthur Whitten Brown.**

Four teams assembled in Newfoundland for the Atlantic bid in mid-1919, turning the attempt on the prize into a race to be ready. Two had made failed attempts when Alcock and Brown took off from an improvised field of marginal length (an anxious moment!), on 14th June 1919.

Bad weather and icing caused loss of instruments and consequent loss of control, recovery on one occasion taking place so low that spray reportedly touched the wings.

After 16 hours and 27 minutes they landed in Ireland. Understandably after that time in an open cockpit they wanted to land quickly, but another motivation was a desire to report a landing before any other claimants who might have followed them. Alas, what had looked to be a smooth field proved to be an Irish bog and the Vimy tipped on its nose.

Alcock did not enjoy his triumph long, for in December of that year he was delivering a Vickers Viking amphibian to France in poor weather when he hit a tree and was killed.

ALDRIN, EDWIN ('Buzz') (1930-) – the second man to walk on the moon. Long before becoming an astronaut, Aldrin wrote a thesis on orbital rendezvous techniques for the Massachusetts Institute of Technology, a prophetic choice of subject.

He became a fighter pilot and flew 66 combats missions in Korea.

In 1966 he flew aboard the *Gemini 12* space mission, setting a spacewalk record for the time of 5 hours, 37 minutes.

He was appointed pilot of the lunar lander *Eagle* for the *Apollo 11* mission, commanded by **Neil Armstrong**. It was reputed that he felt his position on the mission should have entitled him to the privilege of being the first to set foot on the moon.

Nicholas Alkemade made his unintended claim to fame in aviation history while serving as a tail gunner on a Lancaster. He shot down an attacking Junkers 88, but not before his own aircraft was hit and set on fire.

He was unable to reach his parachute through the flames and jumped without it rather than die by fire. His fall was broken by fir trees, then by a snow bank. Astonishingly, not a bone was broken, but his troubles were far from over for the Germans believed from the lack of a parachute that he was a spy.

It was only when he persuaded them to search the wreckage and find the fittings that they accepted his story and spared him from execution. They presented him with a certificate confirming his 18,000 ft (5,500 m) free-fall.

ALLEN, BRYAN (1952-) – flew the first man-powered crossing of the English Channel. **Paul MacCready**, an American glider pilot, designed a man-powered aeroplane in response to a prize offered by industrialist Henry Kremer for a 'figure-of-eight' flight. The prize was originally £5,000 in 1959, but Kremer raised it in stages to £50,000. Many groups had tried for it, and some had flown straight and level, but none had managed to turn their unwieldy machines.

Allen won the Kremer prize in MacCready's *Gossamer Condor* on 23rd August 1977. It was perhaps the first record-breaking aeroplane flight in which the ground crew could run alongside shouting encouragement!

The ever-generous Kremer offered a further £100,000 for a man-powered crossing of the English Channel. MacCready built a new design, the *Gossamer Albatross*, and Allen duly obliged with a crossing in 2 hours, 49 minutes on 12th June 1979 in a prodigious feat of endurance.

Pedal-Power Dream

What could be more enticing than the prospect of taking a modified glider to an open space, pedalling it off the ground and soaring away?

Unfortunately the energy needed to become airborne proved to be at the limit of human ability. It was soon realised that an athlete trained as a pilot had a better chance than the other way round. Allen's channel crossing was similar in effort to cycling up a 1 in 7 hill for nearly three hours!

AMUNDSEN, ROALD (1872-1928) – pioneer of North Polar flights. The Norwegian explorer is best known as the first to sail the North-West Passage and the first to reach the South Pole.

In 1925 he sought to add to his polar achievements by becoming the first to reach the North Pole by air. The expedition ended when one of the two Dornier Wal flying boats was force landed. The crew of the second aircraft landed in dangerous conditions to pick up the occupants.

The following year he tried again, this time using the airship *Norge*. He was, supposedly, just beaten to the Pole by **Richard Byrd**, although there was acrimonious dispute about Byrd's claim. Evidence recently discovered suggests he turned back short of the Pole and that the honour is rightfully Amundsen's after all.

Amundsen disappeared aboard a flying-boat whilst searching for the crew of the airship Italia.

ANTONOV, OLEG (1906-84) – Russian aircraft designer. Antonov built his first glider in 1924, when he was 18. It could hardly have been more different from one of his later designs, his vast An-22 transport of 1965, holder of many payload records and at the time the largest aircraft flying.

He designed a number of pre-war and wartime Russian military types, most of them while working for **Yakovlev**. In 1946 he was able to set up his own organisation, and one of its first products was one of the most unlikely success stories in aviation, the An-2 biplane transport. The rugged single-engine machine looked dated even in 1947 when it first flew, but many thousands were built in Russia, China and Poland, with production continuing into the 1990s.

Antonov concentrated on transports throughout his post-war career, and his designs provided the bulk of Soviet military air-lifting capacity.

ANZANI, ALESSANDRO (1872-1956) – early aero-engine builder. An Italian racing cyclist and motorcycle builder, Anzani moved to Paris and there turned to aero-engine design. His engines had a reputation for being temperamental, a trait supposedly shared with their creator, and were notorious for spewing oil over the hapless pilots.

Their greatest moment came in 1909 when a 23 hp three-cylinder Anzani engine propelled Louis Blériotériot (just!) on his historic flight across the English Channel.

APPLEBY, JOHN (19th cent.) – possibly launched in glider of 1853. **Sir George Cayley** tested a glider in 1849 with a ten-year-old boy aboard. In 1853 (probable date) he installed his coachman aboard and launched him across a valley. The coachman's identity is unconfirmed, but John Appleby is seen as the likely candidate as he was employed in that capacity at the time.

Evidently he was a reluctant pioneer, for on landing he forthwith handed in his notice protesting "I was hired to drive, not to fly."

ARLANDES, MARQUIS D' (1742-1809) – French nobleman aboard man's first ever flight. After the **Montgolfier** brothers had flown unmanned balloons, they started planning a manned ascent. Reputedly (not all historians accept the story), the king of France, Louis XVI, insisted that criminals, whom he evidently considered expendable, should be aboard. In what he considered the unlikely possibility that they survived, they would be pardoned.

A scientist who was closely following the progress of ballooning, **Pilâtre de Rozier**, was so outraged at the prospect of 'vile criminals' taking the glory for man's first venture into the air that he sought help from the Marquis to influence the king. The nobleman did so, on the rash condition that he be aboard the balloon.

Arlandes, Marquis d' *(continued)*

The great moment came, with Pilâtre de Rozier in command, on 21st November 1783, at the Château de Muette in Paris. The pessimism of the King and many others was almost justified, for the open fire repeatedly burned holes in the inflammable fabric of the balloon.

Evidently this was expected, for sponges and buckets were to hand and vigorous action with these kept matters under some semblance of control. The courageous aeronauts landed after a flight of 25 minutes.

ARMSTRONG, NEIL (1930-) – first man on the moon. Armstrong started his flying career as a fighter pilot, recording 78 combat missions in Korea. His flying skills were recognised by his transfer to test flying, the background of most early astronauts.

Once selected for the manned space programme, he demonstrated his exceptional coolness when his Gemini spacecraft started spinning uncontrollably whilst in orbit.

Later he was to survive another close call when he ejected at low level from a test version of the lunar lander which had gone out of control.

He was awarded the supreme privilege of commanding the *Apollo 11* mission, the first to land on the moon. The historic landing was made on 20th July 1969, followed by mankind's first steps on the moon the next day.

It is a fallacy that computers had taken over all the skill of flying on space missions, for Armstrong and **Aldrin's** approach and landing on the lunar surface called for piloting skills of the first order. As they manoeuvred above boulders to find a landing site they were assailed by computer warnings of problems and Mission Control advising ever more anxiously about the fast diminishing fuel reserve, a most stressful set of conditions.

The landing was achieved with minimal fuel remaining, reportedly enough for just 11 more seconds.

After the junketing amid the triumphant return, Armstrong left the space programme and entered private industry. In later years he has almost disappeared from public view.

His piloting ability made him a clear choice to command this most historic of space missions, but he was never as happy handling public relations as some of his colleagues, and in this respect he was not ideal.

The Race for the Moon

Ever since America announced plans to launch a satellite, an unofficial 'space race' developed with the Soviet Union. Early honours went to Russia, with the launch of the first satellite, Sputnik 1, in 1957, followed by the first manned spaceflight in 1961. President Kennedy was outraged and pledged that America would land a man on the moon by the end of the decade.

It was a stupendous feat of organisation which achieved this goal. In the event the Soviet Union never launched a manned moon mission. The Apollo programme was a remarkable triumph, with one tragedy and another near-disaster, but almost as remarkable is the fact that no man has returned to the moon for over twenty years.

Words of Destiny

Neil Armstrong's first words from the surface of the moon were "That's one small step for a man, one giant leap for mankind", or at least that is what he meant to say.

In the excitement of the moment he appears to have forgotten his lines and omitted the 'a', losing the meaning of his carefully chosen sentence.

ARNOLD, GENERAL HARLEY ('Hap') (1886-1950) – Chief of Staff, U.S. Army Air Corps, 1938-1940. Arnold was one of the earliest of aviators, learning to fly in 1911 with none less than **Orville Wright.**

His crucial role in history was lobbying for expansion of the U.S. Army Air Corps (Air Force from 1940) which was to prove so vital to the Allied cause during the Second World War

BACON, ROGER (c.1214-1292) – believed to have been the first to write about manned flight. An English Franciscan monk, Bacon wrote a book, 'Secrets of Art and Nature', around 1250 AD, in which he described 'engines for flying'.

Like so many to come, they were flapping-wing devices (ornithopters). These were to prove a dead end in the quest for flight, but he is believed to have been the first to approach the problems scientifically. Visionary he may have been, but others saw his work differently and he was imprisoned for heresy.

BADER, SIR DOUGLAS (1910-82) – the legendary legless fighter pilot. A year after entering the RAF, Bader misjudged a low-level aerobatic sequence and touched the ground with a wing-tip. Both legs were amputated after the crash.

He was invalided from the service, but on the outbreak of war in 1939, after several attempts, he was accepted once more for flying despite his disability.

He was credited with 22½ victories as a fighter pilot. In 1940 he took command of a squadron, then demoralised after its experiences in France, and he quickly turned it into a potent fighting unit. In 1941 he was awarded leadership of a wing.

In August 1941 he was shot down, or was involved in a collision, and taken prisoner. Improbable as it might seem, he tried to escape and was sent to the notorious Colditz Castle.

He was a strong advocate of **Trafford Leigh-Mallory's** 'big wing' theory, whereby the enemy could be better tackled by forming several squadrons into a wing.

He may even have been the originator of the tactic. He was certainly called upon to explain its merits to a meeting of the Air Staff, despite his then modest rank of Squadron Leader.

Post-war he devoted himself to helping limbless people, as well as his work flying and managing the Shell company's aircraft fleet.

BALBO, ITALO (1896-1940) – leader of long-distance formation flights. As Secretary of State for Air, Balbo promoted the Italian Air Force by leading two impressive formation flights.

The first comprised 14 Savoia-Marchetti SM55A twin-engine flying-boats which he led on a mass flight to Brazil in 1930. Four failed to complete the round trip and five fatalities were incurred, but Balbo was encouraged enough to plan a yet more ambitious journey.

This flight involved 24 improved SM55X flying-boats which set off for the 1933 Chicago World Fair. Apart from one fatal accident, the 'Armada' was a triumph of flying and organisation.

Balbo opposed Italy entering the war on Germany's side. He was shot down over Libya in 1940 by Italian gunners. Supposedly an accident, it is widely believed Mussolini ordered his removal because of their differences over policy towards the war.

BALL, ALBERT (1896-1917) – leading World War One fighter pilot. Ball was credited with 44 victories, although his true total was probably higher as he was modest about his claims – some sources quote 47.

Albert Ball – tranquil on the ground, lethal in the air

He was a quiet man who preferred to fly alone. In contrast to his uncompromising aggression in the air, between sorties he would quietly play the violin, tend a garden he had created, and attend church.

He was killed in 1917 while flying an SE5. Reportedly he was hit by ground fire, but this is uncertain as his last moments were not witnessed. He was awarded a posthumous VC.

Courage above the Trenches

Few of the First World War 'aces' survived, their life expectancy being short because none of the Allied aircrew and few of the Germans carried parachutes. All pilots dreaded fire in the air and some carried pistols with, it was said, 'one round and one purpose'.

BANKS, RODWELL (1898-1985) – engine and fuel specialist. A naval officer in command of motor torpedo boats in the First World War, Banks specialised in engine development after the war, first for marine use and later extended to cars and aircraft.

By developing special fuels to avoid engine 'knocking', Banks played a significant part in the British Schneider Trophy wins between 1927 and 1931. Later, his continued work in this field yielded the 100 Octane fuel which arrived just in time for the Battle of Britain and gave crucial extra speed to RAF fighters.

He became Director of Engine production under Lord Beaverbrook, a most responsible post in the wartime period. Post-war he worked in industry, with Bristol Engines and later as a director of Hawker Siddeley.

BARNWELL, FRANK (1880-1938) – designer for the Bristol Aeroplane Company. Barnwell joined the company, then British and Colonial, in 1911, and was involved in or responsible for Bristol designs for the next 27 years apart from one short break overseas.

His most important First World War design was the Bristol Fighter, or 'Brisfit'. One of the most successful Allied combat aeroplanes, some 5,500 were made.

Barnwell, Frank *(continued)*

In 1935 Barnwell produced a neat twin-engine monoplane, the Bristol 142. It proved 50 mph (80 kph) faster than the best RAF fighters. From it he developed the Blenheim bomber, an advanced machine when it flew in 1937 but already becoming obsolete by the outbreak of war. It served widely in RAF squadrons as a light bomber and as a fighter.

In turn the Blenheim led to the potent Beaufighter which was the best of the famous Bristol twins. Alas, Barnwell never saw the distinguished record of the Beaufighter. In 1938 he built and flew a light aircraft. In contrast to his renown at the drawing board, he was a notoriously poor pilot and he killed himself flying it.

BATTEN, JEAN (1909-82) – long-distance record-breaker of the 1930s. A New Zealander, Jean Batten learned to fly in Britain and bought a second-hand Gipsy Moth for a solo flight to Australia. Her first two attempts ended in forced landings, but the tenacious young woman tried again and in May 1934 reached Australia in record time.

She made the return flight the following year. She suffered an engine failure over the sea and was about to ditch when the engine recovered at worryingly low altitude.

Not content with this achievement, she made further record-breaking flights to South America and New Zealand. For these she flew a Percival Gull, which at least offered the protection of a closed cockpit but still exposed her to relying on a single engine over long over-water stretches.

In her last years she disappeared and appeals were made in aviation magazines to trace her. Eventually it was learned that this great pioneer woman aviator had died in Majorca.

Fighter pilot Raymond Baxter, later a well-known BBC commentator, fired what he called "the most optimistic shot of the war" when he saw a V2 rocket rising in front of him. It would have been a real 'line' if he had destroyed it, but it was just too fast for a realistic chance.

BEAMONT, ROLAND ('Bea') (1920-) – fighter and test pilot. After flying Hurricanes in the Battle of Britain, Beamont spent a nominal 'rest' period testing the Hawker Typhoon and Tempest. The former had a justifiably nasty reputation for structural failures and for its temperamental Napier Sabre engine.

Back on operations, he commanded first a Typhoon squadron and then the first Tempest wing. In 1944 his Tempest was damaged and he was taken prisoner.

After the war he returned to test flying, handling Meteors at Gloster before becoming Chief Test Pilot at English Electric in 1947. He handled the maiden flights of the Canberra (1949), Lightning (as the P1 in 1954) and the TSR2 (1964). He was a notably accomplished display pilot.

After retirement from test flying he handled military sales for BAC, of which English Electric had become a part, and ultimately British Aerospace.

BEAVERBROOK, BARON (William Maxwell Aitken) (1879-1964) – Minister of Aircraft Production during the Second World War. Lord Beaverbrook, the Canadian-born newspaper owner, was appointed Minister of Aircraft Production in 1940 by Churchill. With enormous energy and justifiable ruthlessness he cut out 'red tape' and other bottlenecks to producing the aircraft vital to Britain's survival.

So effective were his measures that the RAF ended the Battle of Britain with more fighters than at the beginning and Britain's output exceeded Germany's. It was the shortage of pilots which was the most critical problem at that time.
His son, Max Aitken, was a distinguished fighter pilot.

BEDFORD, WILLIAM ('Bill') (1920-) – test pilot for Hawker Siddeley. Bill Bedford joined the team testing the Hunter in 1951. He was renowned for his spinning demonstrations including, perhaps uniquely in a jet aircraft, inverted spins at Farnborough Air Shows.

In 1961 he achieved the first transition from vertical to wing-borne flight of the Hawker P1127, the predecessor of the Harrier. The programme was a hazardous one, for this was an entirely new type of piloting. He survived an ejection following an engine nozzle failure, and was lucky to walk away from an ignominious crash at the 1963 Paris Air Show when the nozzles jammed. He retired from test flying in 1967.

BEECH, WALTER (1891-1950) – founder of the Beech company. Having learned to fly in 1914, Beech became an Army flying instructor before turning to earning a living by 'barnstorming'.
In 1924 he formed his own company, Travel Air, which he sold in 1929. He founded the Beech company in 1932, selling a neat business biplane popularly called the Staggerwing because the top wing was mounted aft of the lower one. Later examples were unusual for biplanes in using retractable undercarriages.

Beech, Walter *(continued)*

A notable pre-war success was his twin-engine Beech 18, a monoplane widely used for business transport and wartime military training. Its 33 year production span for long held the record for any aeroplane.
After the war Beech launched his single-engine Bonanza with its distinctive 'V' or butterfly tail. It was a huge success, with over 10,000 made.
The company remains a major builder of business aircraft, now as part of the Raytheon group. Beech himself saw little of the post-war success, for he died in 1950. His widow and former company secretary, Olive, took over as chairman and held the post for 30 years.

BELL, LAWRENCE (1894-1956) – founder of the Bell Aircraft Corporation. After working for Martin then Consolidated, Bell started his own company in 1935. Over his working life he had an astonishing record of innovation, with the first American jet, the first supersonic aircraft, the first jet vertical take-off aeroplane, and the first 'swing-wing' all to his credit. As if this was not enough in the fixed-wing field, he built the first helicopter to be certified too!

Bell built large numbers of fighters in World War Two, notably the Airacobra. In typical Bell fashion he tried a layout unique to him in mounting the engine behind the pilot. The pilot tried not to think about the long shaft spinning between his legs driving the propeller. In fact the performance was not in the top league and the bulk of deliveries went to Russia.

The first American jet aeroplane was the Bell XP-59A which first flew on 1st October 1942. In a sense it was the first jet fighter anywhere, for it was the first to carry guns. Once again, performance was disappointing and only a small production batch was made.

Bell's greatest moment of glory was the world's first supersonic flight in the XS-1, better known by its later designation X-1. Captain **'Chuck' Yeager** dropped from under a B-29 bomber in the rocket research aircraft and reached supersonic speed on 14th October 1947.

Bell was one of the first to see the possibilities of the helicopter, and flew his little Bell 47 in 1945. It received civil certification the next year, the first in the world to do so. He had picked a winner, for it was made for 31 years. Since then Bell has concentrated on helicopters, another notable success being the Bell 204, or military HU-1 'Huey', of which over 10,000 were made.

BELLANCA, GUISEPPE (1886-1960) – designer of efficient light aircraft. A Sicilian who emigrated to the United States in 1910, Bellanca learned to fly in an aircraft he had built himself. His first successful design was his WB-2 of 1926. Bellancas proved to be exceptionally efficient, achieving high speeds and long ranges for their class.

Lindbergh's first choice for his Atlantic flight was a Bellanca, but the designer's business partner imposed unacceptable conditions for a sale. Shortly after Lindbergh's flight, in a Ryan, a Bellanca showed it had the capability for such a distance by making a non-stop flight from America to Germany.

In post-war years Bellanca continued building well-respected light aeroplanes, until he lost control of his company in 1959 and he turned to farming. In a complex series of family and business affairs, there were actually two Bellanca companies for some years.

BENNETT, DONALD (1910-86) – founder of the 'Pathfinders'. An Australian, Bennett started his flying career with the RAAF but came to Britain and transferred to the RAF in 1931. He flew as navigator in the 1934 MacRobertson air race, but was injured in a crash in Syria.

He joined Imperial Airways in 1935 and flew flying-boats. He participated in the trials of the Mercury-Maia composite aircraft, an idea to achieve long range whereby a seaplane took off 'piggy-back' on top of a large flying-boat. He flew the upper part, *Mercury,* and set several long-distance records. One such, in 1938, is considered the first westbound commercial Atlantic aeroplane flight. The performance was impressive enough but it was hardly an economic proposition for regular services.

Bennett was one of the founder members of the Atlantic Ferry Service, the organisation which flew large numbers of desperately-needed aircraft from Canada and the United States to wartime Britain, and ran a return service to take the crews back for more.

He switched to bombing operations in 1941. Among his exploits was leading an attack on the battleship *Tirpitz* in a Norwegian Fjord. His Halifax was hit and he baled out just as it exploded. He reached neutral Sweden on foot.

His experience as a bomber pilot had shown him just how poor the accuracy was on most raids with the limited navigation aids then available. He started the Pathfinder Force in 1942, forming a nucleus of highly trained navigators who dropped markers on the targets to guide the main force. Everything depended on the skill of the Pathfinders, for if their markers landed astray all the bombs would miss the target. After some early setbacks, it generally worked well and the Pathfinder Force flew some 50,000 sorties.

After the war he led the airline British South American Airways. The company suffered a setback with the never fully explained loss of two of its Avro Tudors over the Atlantic. Bennett left the airline in 1948, a year before it was absorbed by BOAC. He formed a new company to work the Berlin airlift and handle general charter work, but suffered yet another Tudor accident. His last aviation enterprise was forming Dart Aircraft to build light aircraft, components, and later hovercraft.

BENOIST, THOMAS - first passenger aircraft service. Most of Tom Benoist's contribution to aviation history has been long forgotten apart from one achievement: he launched the first airline service in the world using aeroplanes (airships had already been so used in Germany).

The service was inaugurated on 21st January 1914 between Tampa and St. Petersburg, using a Benoist flying boat. As only a single passenger could be carried, economic reality brought it to an end after four months. It is a mark of how quickly air travel later progressed that it was just 62 years to the day before the first supersonic passenger services began.

BICKLE, PAUL – holder of the gliding gain of height record. Few record-holders are included in these biographies as the information becomes out of date so quickly, but this record is remarkable for having stood for 35 years.

Bickle flew a Schweizer I-23E sailplane in a 'lee wave' system downwind of the Sierra Nevada mountains in California. The height gain was 42,303 feet (12,894 m). He believed he could have gone higher but for the cold and the limits of his oxygen supply.

Riding the Wave

There are three types of 'lift' commonly used in gliding. These are using the updraught as winds are forced over hills, thermals, and lee waves. These occur downwind of mountains when the air rises and falls in a series of sine waves. They reach many times the height of the mountains and have been used for most recent height records.

So strong are the lee waves in the Sierra Nevada area that the pilot of a Lockheed P-38 fighter, unable to land because of bad weather, is reported to have feathered his engines and soared for an hour until his base opened again.

BISHOP, RONALD (1903-89) – designer of de Havilland aircraft. Bishop joined the company in 1921. The first type for which he was wholly responsible was the Flamingo, a neat twin-engine monoplane produced just before the war and which could have sold in large numbers had hostilities not intervened.

De Havilland turned to military work during the war and proposed a light bomber relying entirely on its speed for safety. To give the best possible performance, no defensive guns would be carried. The idea of omitting guns was sacrilege to many senior officers and there was much opposition, but Bishop's design, the Mosquito, proved one of the finest combat aircraft of the war. He conceived it around two Rolls-Royce Merlin engines, and daringly built it of wood, enabling firms like furniture builders and even coffin makers to contribute towards the war effort. The Mosquito was indeed hard to catch and losses were far lower than for other bombers.

Bishop turned to jets with the Vampire fighter, and to post-war civil work with the elegant Dove light transport.

More radical was the heroic attempt to overtake the American lead in airliners with the world's first jet transport, the Comet. Bishop looked at every possible layout, including tail-first designs, before settling on the beautifully sleek Comet. It flew in 1949 and entered service in 1953. De Havilland looked as though they were on the verge of stunning success, but a series of accidents grounded the early Comet. Improved versions arrived in 1958 but the lead had gone.

Bishop became Deputy Managing Director until his retirement in 1964.

The Designer's Worst Fears

Every designer worries that his creation may fail, perhaps with tragic results. Bishop and his team must have gone through agonies when the Comet ran into trouble. A couple crashed on take-off, but there was worse to come. Another crashed in a tropical storm, then two more disappeared over the Mediterranean.

After an epic of salvage and investigation the cause was traced to metal fatigue, and the lessons learned have made every jet airliner since much safer. The designers might well have given up, but they persevered to ensure that better Comets would fly again, and they did.

BISHOP, WILLIAM (1894-1956) – high-scoring Allied 'ace' of World War One. A Canadian, Bishop was credited with 72 'kills'. For a lone attack on a German airfield he was awarded the VC. Some historians, however, dispute his claimed number of victories, although there is no doubt he genuinely achieved an impressive enough number.

Against all the odds, he survived the war, going on to playing a part in setting up early Canadian airline flying. In the Second World War he helped to organise pilot training. Bishop was one of the few 'aces' to grow old.

BLACKBURN, ROBERT (1885-1955) – pioneer British aircraft builder. After watching **Wilbur Wright** flying in France in 1908, Blackburn built a monoplane of his own in 1909. It failed to achieve true flight, but a more successful design did fly in 1911.

After a few flights, Blackburn concentrated on manufacturing and ceased flying himself.

Blackburn headed his own company for over 40 years. He concentrated upon marine aviation, and Blackburns were noted for a fine series of flying-boats in the interwar years. Unfortunately most of the company's products which saw Second World War service, such as the Roc, Skua, and Botha, were outclassed and much of Blackburn's output was of other firms' designs.

After the war, Blackburn merged with General Aircraft, inheriting their ponderous but practical Beverley transport. The company did not long outlive its founder as an independent concern, becoming part of Hawker Siddeley in 1963.

BLANCHARD, JEAN PIERRE (1750-1809) – first to cross the English Channel by air. Blanchard made his first balloon ascent in 1784, carrying flapping wings in a vain attempt at steering.

The cross-channel flight was made on 7th January 1785 accompanied by his sponsor, Dr. John Jeffries, an American. On departure from Dover he had used a ruse to try and leave without Jeffries to keep all the kudos for himself. It was only after an unseemly scene and intervention of bystanders that Jeffries was able to climb aboard.

Jeffries may soon have wished he had remained behind after all, for over the sea the balloon lost height and the aeronauts had to jettison the flapping wings, an equally futile hand-turned propeller, their coats and even in desperation their trousers. The measures worked and they landed triumphantly, if immodestly, in France.

Blanchard went on to make the first flight in America, on 9th January 1793. He died in 1809 soon after suffering a heart attack in the air.

Playing with Fire

Blanchard's widow was also an accomplished balloonist, making a speciality of launching fireworks displays from the air. This seemingly reckless practice from a hydrogen balloon ended with the almost inevitable result in 1819.

After the envelope exploded she survived landing on a roof but was killed when she fell from there to the ground.

BLÉRIOT, LOUIS (1872-1936) – first to fly an aeroplane across the English Channel. The channel-crossing monoplane was Blériot's eleventh machine. Most of its predecessors had ended in a pile of wreckage, dissipating much of their builder's fortune from his acetylene headlamp business.

Louis Blériot – "Britain is no longer an island"

His first attempt at flight was in 1901 with a flapping-wing contraption, which proved as incapable of leaving the ground as others of its kind. He then teamed up with the **Voisin** brothers for a while. Once he nearly drowned Gabriel Voisin when towing a glider behind a boat – an incident recorded on a very early ciné newsreel.

Several more aeroplanes and as many crashes later, Blériot sought to recoup some of his expenditure by trying for a *Daily Mail* prize of £1,000 for the first aeroplane crossing of the English Channel (it had been done several times by balloon).

Another pilot, **Hubert Latham,** had already tried and come down in the sea, and was now ready for another attempt. Blériot forestalled him by taking off at 4.41 a.m. on 25th July 1909. 35 minutes later he flew through a valley in the cliffs near Dover – his little 23 hp Anzani engine left him somewhat underpowered – and landed heavily but triumphantly near the castle.

Blériot suffered yet another accident in December of that year and gave up flying, no doubt wisely in view of his track record.

BLOCH, MARCEL (1892-1986) – founder of Dassault. Bloch became involved with aircraft building during the First World War, but in the lean post-war period turned to furniture making.

In 1930 he started building light aircraft, followed in 1932 by the more ambitious Bloch 200 bomber. In 1936 he built his Bloch 220 airliner, a neat 16-seat twin-engine monoplane. Bloch might have challenged the Douglas DC-2 with this aircraft, but the company suffered the upheaval of nationalisation soon afterwards.

During the war, Bloch and his brother refused to collaborate with the Germans and Marcel was sent to a concentration camp, narrowly avoiding execution.

Bloch, Marcel *(continued)*

When peace returned he resumed aircraft design, but because the Bloch company was now associated with collaboration he adopted the name Dassault from his brother's resistance code name *Char d'assault* (battle tank). He launched some of the world's finest combat aircraft, including the Ouragan, Mystère, Mirage III and the quite different Mirage F1. Large numbers have been sold throughout the world and have acquitted themselves well in action.

Bloch (or Dassault as he then called himself) branched out into the civil field with a successful series of business jets, but his one major failure was his attempt to enter the airliner market. The Dassault Mercure was designed with too limited a range and the huge investment yielded only 10 sales to the French airline Air Inter.

BOEING, WILLIAM (1881-1956) – founder of the Boeing Airplane Company. Boeing built his first aeroplane, a small flying boat, in 1916. In the same year he formed his company as Pacific Aero Products, adopting the Boeing name in 1917.

In 1927 Bill Boeing won a mail contract between San Francisco and Chicago, forming Boeing Air Transport for the purpose. It was an ancestor of the present-day United Airlines.

Boeing's 247 transport of 1933 is considered a turning-point in airliner design. In an age of wooden biplanes its metal construction and retractable undercarriage set the pattern for years to come. The Boeing company never again achieved pre-eminence in airliners until the end of the piston era, but they were the first American builders of jet airliners and have been the world's leading airliner builders since the beginning of the 1960s.

In the military field many of its bombers were renowned, including the B-17 Flying Fortress, the B-29, and the great jet B-47 and B-52. Bill Boeing retired early from the company, leaving in 1934.

BOELCKE, OSWALD (1891-1916) – pioneer tactician in air warfare. In the early stages of World War One, squadrons tended to comprise a loose grouping of mixed types of aircraft. In 1916 the highly successful German pilot Boelcke proposed the use of specialised fighter squadrons. He was placed in command of the first such unit, inconsistently numbered Jagdstaffel 2. He was allowed to select his pilots for the squadron. **Manfred von Richthofen** was one of his protégés.

Like most combat pilots of the time, his active life was a short one, ending in a collision in October of the same year. He had scored 40 victories.

Boelcke disliked killing his opponents and was respected by the British. After his death, RFC pilots dropped a wreath to "Captain Boelcke, our brave and chivalrous opponent."

BOOTH, RALPH (1895-1969) – airship commander. Booth's first flight in command was unintentional! He was First Officer aboard R-33 at the mooring mast with some of the crew, but not the captain, when a storm engulfed them. The nose was stove in and R-33 broke away. They were blown across the North Sea to Holland, but by fine airmanship and skilful interpretation of the wind pattern around the low pressure area he saved the damaged airship and the crew.

He commanded the R-100, handling much of her testing and making an acclaimed double Atlantic crossing in 1930. The flight was hailed as a triumph, but little mention was made at the time of two near disasters, once in a storm approaching Montreal and the second on the return trip when a helmsman dozed off and almost allowed the airship to descend into the ocean.

The Lighter than Air Dream

The R-100 and ill-fated R-101 were intended to fly regular intercontinental airship services. R-100 never flew again after the R-101 tragedy and was scrapped. She had major problems and the decision was wise.

With hindsight it is hard to see how reliable, economic services could ever have been achieved. To avoid prohibitive loss of gas, the rigid airships cruised below 1,000 feet (not much more than their length), so were at the mercy of all weathers, and overland flights were almost a disaster waiting to happen.

BORMAN, FRANK (1928-) – commanded the first mission around the moon. Borman joined the astronaut team during the Gemini programme, and in the *Gemini 7* spacecraft participated in the first space rendezvous. With James Lovell and William Anders he took *Apollo 8* into orbit around the moon in December 1968, a mission which was in many ways as much a step forward as the moon landing eight months later.

While Borman and his colleagues were the first ever to look upon the far side of the moon, there was a tense wait on earth while they were out of radio contact on their first orbit. Relief was felt all over the world when Borman's voice came through reading passages from Genesis. After retiring as an astronaut, Borman became president of Eastern Airlines.

BRABAZON – ***see Moore-Brabazon***

BRANCKER, SIR SEFTON (1877-1930) – flamboyant Director of Civil Aviation in the 1920s. Formerly a soldier with service going back to the Boer War, Brancker promoted all aspects of flying with enormous energy after his appointment as Director of Civil Aviation in 1922.

Among his achievements was his scheme to assist flying clubs to buy their own aircraft, which did much to launch the de Havilland Moth series. Not content with desk-bound management alone, he took part in a number of long-distant flights himself. His instantly recognisable monocled figure became a familiar sight at airfields throughout Britain and abroad.

He did much to support the struggling airlines too, encouraging passenger services both by aeroplane and airship. It was typical of his enthusiasm that he wished to be aboard the inaugural service of the airship R-101, and he was among those killed when it struck high ground near Beauvais in October 1930.

BRANSON, RICHARD (1950-) – founder of Virgin Atlantic Airlines and balloonist. The founder of the Virgin music empire was encouraged to form an airline by businessman Randolph Fields in 1983. After some reluctance, due to his total lack of experience with aviation, Branson agreed and a transatlantic service was started in 1984 with a single Boeing 747. It was aimed to some extent at filling the gap left by the demise of **Freddie Laker's** *Skytrain.*

The partnership with Fields was short-lived, but Virgin Atlantic Airlines proved much more durable, winning several awards for its service and expanding more-or-less continuously. This has been despite some unethical competitive practices by major competitors, the so-called 'dirty tricks' campaign by British Airways.

In 1987 Branson attempted the first balloon crossing of the North Atlantic, with Per Lindstrand. They achieved the distance but ditched in the sea off Londonderry and did not quite qualify for their goal.

In 1991 the same crew made the first Pacific crossing by balloon, flying from Japan to Canada.

BRAUN, WERNHER VON (1912-77) – rocket pioneer. Von Braun started experiments with liquid-fuelled rockets in 1930. During the war he led the development of the A4 missile, later known as the V2. While technically a triumph, it was far too inaccurate to have any real military value and could only be aimed at large cities, where its arrival without warning was indeed terrifying.
After the war, he settled in the United States and led the team which made the Redstone rocket which launched America's first satellite, on 31st January 1958.

He followed up this success by playing a major role in designing the great Saturn rockets for the moon landings. He finally left the American space programme in 1972.

The V2 Boomerang

For all its technical brilliance, the V2 soaked up resources far out of proportion to its value as a weapon, so much so that Albert Speer, the German Armaments Minister, later admitted it had damaged Germany more than the Allies because of its colossal cost.

Von Braun's own attitude to the V2 programme is still a matter for debate. It is believed his heart was in spaceflight rather than weapons, but how much did he know about or turn a blind eye to the quite dreadful slave-labour camps, in which three men died for every missile produced?

BREGUET, LOUIS (1880-1955) – helicopter pioneer and aircraft builder. Breguet built a helicopter in 1907 which lifted off the ground (just!). It is not normally counted as the first helicopter flight as men kept it steady with poles.

However, becoming airborne in a helicopter is one thing, making it go in the desired direction is another. Breguet realised that the problems of control were beyond him and turned to aeroplanes, forming one of the major French manufacturers.

The most notable flight by any of his aircraft was the first flight from France to America by the Frenchmen Costes and Bellontes in their Breguet 19 *Point d'Interrogation* in 1929.

BRIS, JEAN-MARIE LE (1808-72) – flew a glider in the 1850s. The glider, shaped like an albatross following his observations of birds whilst he was a sea captain (an excellent choice – the albatross is an outstanding soaring bird), was towed like a kite behind a horse.

It reportedly reached over 300 ft (100 m), but on one occasion the rope wrapped itself round the coachman and carried him aloft. Evidently the job of coachman had its hazards – *see* **John Appleby** – but happily he was not seriously hurt. This mishap and another when he broke a leg notwithstanding, le Bris showed a good grasp of the fundamentals of flight.

BRISTOW, ALAN (1923-) – founder of Bristow Helicopters. After wartime service with the Fleet Air Arm, Bristow first became involved with helicopters during a brief spell as a test pilot with Westland. A strong personality, he soon fell out with them and left to handle flying jobs anywhere they turned up around the world, including whale spotting over Antarctica (still respectable then!)

He formed Bristow Helicopters in 1960, leading it to become one of the largest companies in its field in the world. He remained chairman until 1985.

BRITTEN, JOHN (1928-77) – co-founder of Britten-Norman Aircraft. With partner Desmond Norman the company was making crop-spraying equipment in the 1960s when they found there was no real replacement for the de Havilland Rapide biplane light transport. They were not the type of men to bewail the lack of the right equipment so they made their own.

Their own design was the BN-2 Islander which first flew in 1965. It was an extremely rugged machine which soon sold to remote areas all over the world, over 1,000 having been delivered.

Unfortunately sales success was not matched financially and the company passed through various changes of ownership before passing to Pilatus shortly after Britten's early death.

A three-engine enlarged version of the Islander was also built, the Trislander. The third engine was ingeniously mounted at the top of the fin, giving the aircraft an unusual appearance. One observer on seeing it for the first time exclaimed "Oh look, two aeroplanes mating!"

BROWN, SIR ARTHUR WHITTEN (1886-1948) – navigator on first non-stop Atlantic flight. Brown (sometimes written as Whitten-Brown) flew as an observer during the First World War and became a prisoner after being shot down and wounded in the leg in 1915.

During the Atlantic flight with **John Alcock** in 1919, severe icing occurred in cloud and Brown repeatedly had to stand up in the open cockpit to clear ice from the instruments and their measuring points. It is a tribute to his skill that their landfall in Ireland was commendably close to their planned track.

Brown, like his pilot, was knighted for the achievement. He ceased flying and spent the rest of his career in industry. His son, also Arthur, was shot down on D-Day, after which the father became almost a recluse. In 1948 he took his own life, a tragic end after a glorious contribution to the progress of flight.

BULMAN, PAUL ('George') (1896-1963) – test pilot and first to fly the Hurricane. After service with the RFC during the First World War - he would not divulge the number of his victories - 'George' Bulman became one of the outstanding test pilots of his day, first at Farnborough, then at Hawkers.

He made the maiden flight of the Hurricane, wearing his traditional trilby, on 6th November 1935. He played an important part in the development of this fighter which was to prove so crucial in the years to come. Almost bald, and far from the popular image of a test pilot in appearance, he flew for Hawkers throughout the war years, retiring from flying in 1945.

One American president has seen active service in the air – George Bush. He was a wartime carrier pilot, and was once shot down. Service from carriers was one of the most hazardous forms of wartime flying, for the chances of rescue from the sea in a combat area were slender.

BYRD, RICHARD (1888-1957) – first to fly over both geographic poles. A naval aviator, Byrd reached the North Pole with Floyd Bennett in a Fokker trimotor on 9th May 1926, just beating **Amundsen** who was heading for the same objective. Some observers doubted whether Byrd actually reached the pole, a controversy which was never really resolved.

In 1927 he made a bid for the first flight from America to France, but an accident put him out of the running and he generously let **Charles Lindbergh** fly from his airfield for his historic flight. Byrd did later attempt the same flight but was forced to ditch in the sea off France.

The South Pole was reached with Bernt Balchen and two others in a Ford Trimotor on 28th November 1929. In 1956 he flew to the South Pole again, this time in a U.S. Navy R4D (naval Douglas DC-3).

CALTHROP, EVERARD (1857-1927) – inventor of the *Guardian Angel* parachute. Oddly enough for a parachute pioneer, Calthrop's background was in railway engineering. The *Guardian Angel* was developed in 1914 and demonstrated in live drops the following year. It worked well, but its use was opposed by senior ground-based officers and it was never used by the British services other than for dropping agents. It was used successfully by Italian airmen.

Calthrop died after spending £20,000 on development, understandably bitter that thousands of lives which his creation could have saved had been lost.

The Parachute Scandal

The resistance of service chiefs of staff to the provision of parachutes was one of the scandals of World War One. Even pilots who wished to buy their own from Calthrop were forbidden to do so.

The objection usually made was that aircrew might bale out rather than face combat. In fact the reverse would have been more likely, as pilots would have been more inclined to press home an attack if a means of escape was available. Moreover there was the obvious point that aircrew saved by parachute were available for further service.

As it was, many thousands were sent unnecessarily to terrible deaths.

CAMERON, DONALD (1939-) – maker of hot-air balloons and airships. Cameron was one of the first to re-introduce hot-air ballooning to Britain, after a long absence since the 18th century. He made several record-breaking flights, and just failed in a bid to make the first Atlantic crossing.

His company specialises in making exotic shaped balloons, often in the shape of company emblems for advertising purposes.

In 1973 he flew the world's first hot-air airship.

CAMM, SIR SYDNEY (1893-1966) – Chief Designer for Hawkers. Camm's interest in flying was triggered by seeing plans to make a model Wright biplane. On such small incidents do choices of career often depend. After a short period with Martinsyde, he joined Hawker in 1923.

His ability was quickly recognised, for he became Chief Designer in 1925. Early designs included his elegant series of military biplanes such as the Fury and Hart.

His most famous design was the Hurricane, first flown in 1935. A robust fighter, which it is believed never suffered a single structural failure unless damaged in action, it also had the useful attribute of being steady when the guns were fired. Far more Hurricanes than Spitfires flew in the Battle of Britain and it was a vital element in the nation's defence.

His later wartime designs, the Typhoon and Tempest, gave a good deal of trouble in development but were eventually turned into effective ground attack aircraft and fighters.

After the war Camm turned to jets, first with the straight-wing Sea Hawk and then with the swept-wing Hunter, surely one of the most pleasing-looking combat aeroplanes ever built. It flew with the British services for over 40 years.

His supersonic successor to the Hunter, the Hawker P1121, was partly built when cancelled by the government. Camm described it as possibly the greatest missed opportunity in British aviation.

The cancellation ensured that French Mirages and American F-16s would fly with the airforces of the world instead of British products.

In the 1960s he became interested in vertical take-off, and with **Stanley Hooker** of Bristol Siddeley engines conceived the Harrier. It is a tribute to Camm that for more than a quarter of a century it has been the only successful aircraft in its class. Like his Hurricane, it also played a vital role in winning a war – the recovery of the Falklands in 1982 relied heavily on the unique capability of the Harrier.

Single-Minded Camm!

Every one of Camm's designs were single-engine aircraft. They are more agile and cheaper. Some prefer twin-engine fighters on safety grounds, but statistically Camm was right - the risk to aircrew is about the same whether there is one engine or two.

Most types of accidents which kill, such as collisions or hitting high ground, are unrelated to power loss. With piston-engine aeroplanes there was a good chance of a successful forced landing, and in jets a power loss generally gives time to eject.

CAMPBELL BLACK, Tom (1899-1936) – racing and trail-blazing pilot. After serving in the RNAS and RAF in the First World War, Campbell Black became a successful long-distance and racing pilot. With Charles Scott he flew the de Havilland Comet which won the MacRobertson Race from Britain to Australia in 1934.

In 1935 he flew another Comet in a bid to take the record to South Africa, but had to bale out with his co-pilot over the Sudan due to propeller trouble.

Such flying was hazardous and the casualty rate was high - indeed by today's standards it might be seen as scandalously so. It was ironic therefore that Campbell Black was killed, not by the obvious dangers of racing, but in a ground collision at Liverpool.

CAPPER, SIR JOHN (1861-1955) – encouraged early military aviation in Britain. It might have been supposed that senior army officers would instantly have grasped the military possibilities of the aeroplane, at the least in reconnaissance for checking on enemy positions and movements. Strangely enough this was not so in the early years of the century, even though heavier-than-air flight was becoming a reality.

It was largely due to the persuasion of Colonel Capper and his predecessor, **James Templer**, that the army belatedly recognised the potential of aviation.

Capper had been involved with ballooning as far back as 1883, and he took command of the army balloon establishment at Farnborough in 1906. He took up ballooning himself to understand the needs of lighter-than-air flying, and flew in races.

In 1907 he piloted Britain's first airship, *Nulli Secundus,* and after a few flights flew the craft to London, landing at the Crystal Palace. There was great public acclaim, but not for the sequel: before Capper could fly her away again the wind rose and the envelope had to be slit with a knife to prevent the airship blowing away.

Capper had the airship rebuilt as *Nulli Secundus II* ('Second to None the Second!'), but he was no designer and the new version proved less satisfactory and was soon grounded.

On the heavier-than-air side, he visited the **Wrights** at his own expense and encouraged **Samuel Cody** in his experiments with kites, and later with his first aeroplane flight in Britain.

CAPRONI, GIANNI (1886-1957) – pioneer Italian pilot and constructor. Caproni flew his first aircraft in 1910 at Malpensa, later the site of one of Milan's airports. In 1913 he flew from Milan to Rome, an impressive flight for the time. In the war years he was best known for his bombers.

He built many fine aircraft but two were unusual. The first was an extraordinary flying boat, the Ca.60 built in 1921 to carry 100 passengers to be borne aloft on no less than nine wings, in three banks of three. Dubbed the *Capronissimo,* it crashed on its first flight.

The other noteworthy machine was the Caproni-Campini CC2, which was claimed to be the second jet aircraft in the world when it flew on 28th August 1940.

Built in association with Secondo Campini, it looked like a turbojet powered aeroplane but inside was a three-stage fan driven by a 900 hp piston engine. No gas turbine was involved and it was really just a ducted propeller. Performance was poor, but there were features pointing to the future, such as a form of afterburning and variable nozzles.

CARTER, GEORGE (1889-1969) – designer of the first British jet aircraft. Having worked successively for Sopwith, Hawker, Shorts and de Havilland, Carter became Chief Designer at Gloster in 1936.

In association with **Frank Whittle** he designed the first British jet aeroplane, the Gloster E28/39, which made its maiden flight on 15th May 1941.

He followed the E28/39 with the first operational British jet, the Meteor. He selected a twin-engine design because of the low power of the early engines. After frustrating delays to the engine programme, Meteors entered service in July 1944, probably just beating the Messerschmitt 262 as the first jet in service. Not all historians agree on which was first as it is not always clear-cut when an aircraft is truly 'in service'.

He remained at Glosters in various posts until retirement in 1958.

CAYLEY, SIR GEORGE (1773-1857) – pioneer of the theory of flight. Often called 'The Father of Aeronautics', George Cayley drew fixed-wing (i.e. non-flapping) designs as early as 1799. A silver disc he engraved in that year shows the earliest known diagram of what is recognisably an aeroplane.

George Cayley

Visionary 'Father of Aeronautics'

He understood the need for curved aerofoil sections, and calculated the lift and power needed for flight mathematically. He built a model glider in 1804, the first flyable model in history, but naturally there was no engine then available for him to advance to powered flight.

He was far ahead of his time – reputedly he devoted so much of his time to science to stay away from his mercurial wife.

After an interval in which he concentrated on other interests, he built another glider in 1849. This was a triplane which successfully carried a ten-year-old boy.

In 1853 he returned to a monoplane glider in which he launched his protesting coachman across a valley. A replica of this glider was flown in 1973 demonstrating its ability to fly, although with such frightening lack of control that perhaps Cayley's coachman had some cause for alarm!

Nevertheless those early trials were a remarkable achievement and his work did much to inspire pioneers like **Pilcher** and **Lilienthal**, who gave the impetus to the **Wright Brothers.**

CESSNA, CLYDE (1880-1954) – founder of the Cessna company. Probably more people have learned to fly on Cessnas than any other make of aircraft. It all started modestly when Cessna watched a flying demonstration in 1911 which inspired him to buy a Blériot, which he altered so radically that it became almost his own design.

Like many others of his time he taught himself to fly by a process of trial and error, indeed so much of the latter that he reputedly suffered 13 accidents before mastering flying skills. At last he succeeded, to such an extent that he became a 'barnstorming' display pilot.

In 1916 he built a design of his own, the Cessna Comet, which set a United States speed record of 125 mph (200 kph) in 1917.

For a time he returned to his original career of farming, before joining **Walter Beech** in aircraft building but, wishing to advance into monoplane design against Beech's conservative preference for biplanes, he set up his own company in 1927.

Cessnas established a fine reputation in racing and amongst private pilots. For a time the company closed in the Depression, but reformed to become one of the world's leading light aeroplane builders. Clyde Cessna retired in 1937. His firm reigned supreme among light aircraft suppliers in post-war years, then left this field for a time to concentrate on business aviation because of potential litigation costs, before once again returning to the private pilot market.

CHADWICK, ROY (1893-1947) – designer of Avro aircraft. Chadwick joined **A.V. Roe** in 1911 and designed such classics as the Avro 504 (with **Roe** himself), the Anson and Lancaster.

The Avro 504 biplane was a principal trainer throughout World War One and well into the 1930s. A few were still employed on special duties with the RAF in 1940.

The twin-engine Anson first flew in 1935. It was the first monoplane and the first aircraft with retractable undercarriage to see RAF service. It was used for maritime patrol, communications, and for twin-engine training. Almost 11,000 were made and the RAF flew the 'Faithful Annie' until 1968.

The Lancaster, for all its later fame, arose from a failure. Chadwick originally designed it to meet a ministry specification as a twin, with 24-cylinder Rolls-Royce Vulture engines. The Vulture was one of the engine maker's few failures, and its problems proved so intractable that **Ernest Hives** ordered it to be abandoned and proposed to Chadwick that four Merlins be used in the Manchester instead.

Chadwick adapted the by then unpopular Manchester, and to break the association named the new version the Lancaster. It proved the best heavy night bomber of the war. Among the exploits of its crews were the famous Dambusters raid and the sinking of the battleship *Tirpitz.*

After the war he turned to airliner design with the Tudor. With his reputation, great hopes were raised for the airliner, but it was not to be and in the end it killed him.

The first version was uneconomical with only 12 passengers, so the stretched 60-passenger Tudor 2 was built. After some alterations it took off on a test flight and immediately banked into the ground, killing Chadwick and test pilot Bill Thorn.

Learning by Mistakes

Safety in the air is improved by learning from accidents and incidents. The Tudor accident was caused by the ailerons being rigged in reverse. The lesson was to make it physically impossible to connect controls incorrectly.

Likewise after a non-return fuel valve was fitted the wrong way round in a Beverley, causing a tragic crash, it became a design standard that such components can only be fitted the correct way.

Chadwick knew well how Ministry minds worked. In the prototype Lancaster he had fitted one more crew seat than specified, which the Ministry men constantly asked him to remove.

When asked by a friend why he upset them by not complying, he explained that by offering them a trivial complaint it would keep them from worrying him about weightier problems.

CHANUTE, OCTAVE (1832-1910) – made early gliding experiments and encouraged the Wrights. Chanute conducted about 1,000 glides, starting in 1896 when he was aged 64. Others flew the gliders under his auspices, due to his age.

His experience as a bridge engineer led to him adopting a biplane layout using the Pratt Truss for strong bracing. It was largely his influence which persuaded the Wright Brothers to follow a similar approach.

His publication of *Progress in Flying Machines* later in 1896 did much to stimulate other pioneers, including the Wrights to whom he gave friendship and encouragement. He even sent two of his helpers and pilots to assist them at Kitty Hawk, and while that 'help' was not always wholly welcome they did acknowledge his contribution to their success.

CHARLES, JACQUES (1746-1823) – pioneer of the hydrogen balloon. Following experiments with unmanned balloons – one had been hacked to pieces by terrified villagers after it landed – physicist Professor Charles built a manned hydrogen balloon with two brothers, Ainé and Cadet Robert. The brothers had devised a means of making silk reasonably gas-tight.

Charles made his ascent with Ainé Robert on 1st December 1783, just 10 days after **Pilâtre de Rozier's** historic flight in a 'Montgolfière'. They started from the Tuileries Gardens in Paris, with reportedly 400,000 spectators (if accurate, half the population of Paris at the time). Charles made a perfect landing 27 miles (43 km) away. He then took off again alone, but perhaps because he underestimated the effect of the lower weight he shot up to 10,000 feet (3,000m) before regaining control and landing safely.

Lighter than Air

Charles never flew again - perhaps that second ascent had frightened him – but the 'Charlière' hydrogen balloon eclipsed the 'Montgolfière' for a long time to come. For a time, town gas could be used, but later a heavier coal gas made this impossible.

In the 1960s it was the turn of the hot-air balloon to make a come-back, using modern propane burners and cutting the cost of ballooning to a fraction of that needed with hydrogen.

CHESHIRE, LEONARD (1917-92) – bomber pilot and founder of the Cheshire Homes. Having completed two 'tours' of operations, Cheshire had reached the rank of Group Captain, at 25 the youngest in the RAF. Several times he had shown remarkable courage, including one incident when much of the side of his Whitley was blown out - he continued on to bomb the target! He accepted a drop in rank to lead 617 Squadron, the 'Dambusters'.

With 617 Squadron he pioneered low-level marking of targets using Mosquitos and later Mustangs. He completed 103 operations and was awarded the VC for courage over a long series of missions rather than one incident. It was the only air VC to be so won in that war.

He was chosen to accompany the second atomic bomb mission, which attacked Nagasaki. The experience left a profound impression on him, but his later research into actions of the Japanese government and Emperor convinced him that the attacks were justified in ending the war.

He founded the Cheshire Homes in 1948 after seeing the plight of sick servicemen who were unable to support themselves, and he dedicated himself to this work for the rest of his life.

CHICHESTER, SIR FRANCIS (1901-72) – long-distance solo pilot and yachtsman. Although better known for his sailing exploits, including the first solo round-the-world voyage in 1966-7, he also made several noteworthy pre-war solo flights.

In 1929 he flew solo from Britain to Australia in a Gipsy Moth, later naming his yachts after the type, and made the first solo crossing of the Tasman Sea. A number of other fine flights followed but they ended with a crash on an intended round-the-world flight.

The Deliberate Error

Chichester devised the idea of 'deliberate error' navigation. Instead of aiming directly towards a small objective such as an island and risk missing it due to wind drift or compass error, he would fly a few degrees away from the direct course. When he should have been abeam the destination, he would turn through around 90 degrees until he reached it.

It was particularly valuable when seeking a town on a river or coastline, where otherwise a pilot could turn the wrong way, away from the objective. The technique became widely used, and was taught by the RAF, until replaced by modern technology with inertial navigation systems and satellite location.

CIERVA, JUAN DE LA (1895-1936) – autogyro pioneer. When a three-engine biplane designed by Cierva crashed, he turned towards rotary-winged flight to avoid the hazards of high-speed take-offs and landings. At first he suffered several crashes until he realised that forward movement gave more lift to the advancing blade than to the retreating one. Cierva made the crucial step of introducing hinged blades.

He built the world's first successful autogyro, which was flown by Alejandro Spencer on 9th January 1923 near Madrid.

He came to Britain in 1926 and produced a number of reasonably successful autogyros which went into production in small quantities. The first rotary-winged English Channel crossing was made by Cierva on 18th September 1928, when he flew a C-8 Autogiro from Croydon to le Bourget, Paris. In World War Two some were used for calibration work.

Cierva was killed in December 1936 when a KLM DC-2 crashed on take-off, a poignant tragedy for a man who had dedicated years to avoiding precisely this sort of hazard.

Cierva's Conception

Unlike a helicopter, an autogyro's blades are not driven in flight. The engine is used solely to drive a propeller, while the rotors are spun by the autogyro's forward speed.

It is a much simpler machine than a helicopter and Cierva was wise to opt for it with the technology of the day. Starting was a rather comical affair on early machines – helpers wound a rope round the rotor shaft to spin it, whereupon the pilot taxied rapidly over the field to reach enough rotor speed for take-off. Later versions could engage the engine briefly for a 'jump' take-off, but they could not hover.

The spelling 'autogiro', as opposed to the generic 'autogyro', was a proprietary name used by Cierva.

CLARKE, ARTHUR (1917-) – first to propose communications satellites. The science fiction author wrote a paper on the use of artificial satellites for communications in 1945, a revolutionary idea at the time. It came to pass 15 years later with the launch of the *Echo* satellite.

Nowadays technology like satellite television seems so commonplace that it is easy to forget just how far-sighted his ideas were. Clarke's vision was a major step into the space age.

CLOUSTON, ARTHUR (1908-84) – racing pilot and record-breaker. A New Zealander, Clouston made a number of record-breaking flights to Australia, New Zealand and Africa. He took part in several long-distance races but was generally unlucky with technical problems.

In the 1930s a Jewish syndicate offered him a million pounds to bomb Hitler. How would history have changed had he done so?

As a test pilot he handled some highly dangerous trials involving deliberately flying into wires to find ways of protecting aircraft against barrage balloon cables.

He flew with Coastal Command during the war, and was later Commandant of the Empire Test Pilots' School.

In the pre-war years the public followed the exploits of leading pilots much as they do pop stars and media celebrities today. Vast crowds turned out to watch the start of races, even in the early hours of the morning, and to see their heroes return. On a record flight to his native New Zealand, Arthur Clouston arrived over his home town and was worried to find it deserted. He wondered what disaster had happened until he reached the airport - almost every citizen was there to see his arrival.

CLYDESDALE, MARQUESS of – led the first flight over Everest. The flight was made on 3rd April 1933, using two Westland biplanes equipped with supercharged Bristol Pegasus engines. Keeping the cameras (and men!) from freezing was difficult, but the task was accomplished and excellent film was taken, which happily still exists.

COBHAM, SIR ALAN (1894-1973) – record-breaker and pioneer of in-flight refuelling. Cobham joined the RFC in 1917 after early military service in the veterinary corps. After the war he gave joy-riding flights, then joined de Havilland in 1921. There he ran their hire service, or what would be called air taxi work now, and handled sales and test flying.

He made many notable long-distance flights, including return flights to Cape Town in 1925-6 and Australia in 1926. This last flight was marred when a Bedouin fired at his DH50 aeroplane and by a freak chance his mechanic, A.B. Elliott, was killed.

In 1927-8 he commanded an epic 23,000 mile (30,000 km) flight round Africa in a Short Singapore flying boat.

In 1934 he formed National Aviation Displays which toured the country giving 'barnstorming' displays and joyrides, and generally promoting air-mindedness. How many later distinguished wartime pilots were inspired to start a flying career by Alan Cobham's displays?

He saw the value of in-flight refuelling after learning of pioneering trials by Captain Smith and Lieutenant Richter in America, and Mrs Victor Bruce in Britain in 1932. His own trials were made in an Airspeed Courier, refuelled by a Handley Page W.10 airliner.

These involved tricky flying in propeller driven aircraft, as vividly seen in one incident when the fuel hose caught between the aileron and wing, putting the Courier into a spin. Despite this, the technique was proven, and it was even used on a handful of airline flights just before and after the war.

He set up his own company to exploit the technique, Flight Refuelling, and he sold his system throughout the world, including the US Navy amongst his customers.

Test pilots in the 1920s were a hardy breed, facing all weathers in open cockpits, but one winter sales tour to Europe in a small de Havilland aeroplane, the DH53, was too much even for Alan Cobham.

The machine proved grossly underpowered, and struggling against a bitter cold headwind the last straw was being overtaken by a Belgian goods train! In disgust he abandoned the DH53 in a field. The next morning the snow was so deep they had trouble finding it.

CODY, SAMUEL (c.1861-1913) – first person to fly an aeroplane in Britain. American born, much of Cody's early history is uncertain. In fact Cody is believed to have been an adopted name. He earned his living as a cowboy, 'bronco-buster', and gold prospector.

In 1890 he came to Britain and toured the country as an entertainer giving trick-riding and shooting displays. He was not, however, 'Buffalo Bill' Cody, with whom he is often confused, although he may have emulated the older man.

Samuel Cody – Swashbuckling American pioneer of British flying

In 1900 he started developing kites for army observation work, and eventually he was launching whole strings of kites which lifted men safely and remarkably steadily.

He built British Army Aeroplane Number 1, which flew on 16th October 1908 on Laffans Plain, later the site of Farnborough airfield. This was the first aircraft flight in Britain. A tree to which he tied his machine for engine runs was preserved as 'Cody's Tree', although today it owes more to fibreglass than to nature.

The flight might have been earlier had not the engine also been used in the first British airship, the Nulli Secundus. Cody was responsible for the propulsion system and procured the Antoinette engine used in both craft.

Cody developed further biplanes, and it was one of these which broke up causing his death.

No Paperwork Please!

Cody was a man of remarkable talents, but strangely enough he was unable to read or write. Some historians have doubted this, but it was well testified by his contemporaries, who confirmed that he even had difficulty writing a signature.

COLLINS, MICHAEL (1930-) – *Apollo 11* Command Module pilot. Collins was the third crew member of the first moon-landing mission, but it was his job to remain in lunar orbit while his colleagues walked on the moon. His role was just as vital to the success of the mission, but because he never set foot on the surface his name is now less well known.

CORNU, PAUL (1881-1944) – made the first helicopter flight. A Frenchman, Cornu just became airborne in a twin-rotor helicopter on 13th November 1907 at Lisieux. While technically a 'first' in aviation history, it was not a practical machine as Cornu had no real means of controlling it.

He was killed in the Second World War by a shell-burst.

Douglas Corrigan took off from New York on 17th July 1938 to fly to California, or so he claimed. He landed in Ireland! He claimed he had misread his compass, but others suspected it was not unconnected with avoiding a ban on attempting an Atlantic flight.

Thereafter he was always called 'Wrong-way Corrigan', but for the rest of his days he stuck to his story.

COTTON, SIDNEY (1894-1969) – pioneer of photographic reconnaissance. From his experience of enduring the bitter cold in open-cockpit biplanes of the RNAS in World War One, Australian-born Cotton developed the heavily insulated Sidcot suit (**Sid**ney **Cot**ton). He could have made a fortune from it, but believed it wrong to profit from the country's needs in wartime.

In the 1920s he flew airmail and survey flights in Canada, with some long-distance air racing to add some spice for an adventurous mind. The survey work led to an interest in aerial photography which in turn prompted the idea of taking clandestine pictures over Germany when war seemed probable. He equipped a Lockheed 12A with hidden cameras and flew it over numerous German military installations, on occasions with senior Luftwaffe officers on board! Even in those last months of peace, the consequences if the cameras had been found would have been dire. He continued so close to the outbreak of war that he risked being interned.

He turned over his facilities to the RAF once war was declared and headed the photographic reconnaissance unit. Sadly he was replaced as head of the organisation he had founded in 1940. An independent-minded man, he found the constraints of service life difficult.

Ellen Church made her modest contribution to history when she boarded a United Airlines Boeing 80A in 1930 - she was the first ever air stewardess. Until well into the 1960s most airlines employed unmarried girls only, but doubtless such a policy would incur wrath today.

COURTNEY, FRANK (1894-1982) – freelance test pilot. A pupil of **Claude Graham-White,** Courtney joined the RFC, after first being turned down due to poor eyesight. After being wounded he transferred to test flying.

He soon established himself as one of the leading freelance test pilots, flying for companies which were too small to warrant employing a full-time pilot. He made over 100 maiden flights.

A tall, easily recognised figure in pinz-nez glasses, the press inflicted on him one of those names of which they are so fond but which make everyone else cringe, in his case 'The Man with the Magic Hands'.

His inevitable share of mishaps included an instance when he noticed the elevator controls becoming slack on an Armstrong Whitworth Sinaia bomber. He managed to land, whereupon the whole fuselage collapsed! Some years later, when testing a Saro flying boat, it was the turn of the wing to collapse on landing.

An Atlantic flight attempt in 1928 ended in the ocean and he was lucky to survive after waiting 24 hours for rescue.

Whilst turning his hand to autogyros for **Cierva,** he pointed out certain weaknesses in the rotor blade design. The Spaniard did not take readily to advice and Courtney was fortunate to escape with his life when he was injured following a rotor blade failure. When Cierva still declined to change his ways Courtney parted company from him.

By 1928 most companies had taken on their own test pilots and Courtney, finding work drying up, left for America, although he returned on occasion for flying assignments. He worked for Curtiss-Wright and later Convair, also handling consultancy jobs.

Test pilot Carl Cover may have saved the Douglas company and changed the course of air transport history when he handled the maiden flight of the Douglas DC-1 on 1st July 1933. Immediately after take-off both engines cut. They restarted when he pushed the nose down, but the process repeated itself every time he tried to raise the nose.

It is a tribute to his skill that he completed a safe circuit. The fault was caused by an assembly error in the fuel system. Had the prototype crashed, the then small Douglas company might well not have survived.

He also handled the historic first flight of the DC-3 on 17th December 1935.

CUNNINGHAM, JOHN (1917-) – night-fighter and test pilot. Whilst flying night-fighters, Cunningham became the first RAF squadron pilot to shoot down an enemy aircraft at night using airborne radar. This was on 20th November 1940, flying a Beaufighter with John Phillipson as radar operator. He was credited with 20 victories, and his success led to his popular (although not with him) appellation 'Cats-eyes Cunningham'.

After the war he returned to de Havilland, which he had joined as a test pilot in 1938 before his RAF service. He became Chief Test Pilot in 1946. Among his maiden flights were those of the Comet (1949), DH110 (1951) and Trident (1964). His test flying and route proving on the Comet made him a household name.

On 27th February 1948 he set a world height record of 59,492 feet (18,000m) in a Vampire with a Ghost engine.

His test flying career was marred towards the end by an unlucky but tragic accident. He was demonstrating a HS125 business jet when it struck a flock of birds on take-off and both engines lost power. He landed without injury to anyone on board but most unfortunately ran across a road at the moment a car taking children to school was passing, killing the occupants. There was nothing he could have done to avoid the accident.

CURTISS, GLENN (1878-1930) – one of the first aeroplane builders and pilots. Curtiss was a racing cyclist, like a remarkable number of the pioneer aviators. Soon he progressed to motorcycles and in 1907 he took the world land speed record on a motorcycle and engine of his own design.

His interest in flight started when an aeroplane syndicate led by Alexander Graham Bell of telephone fame asked him to make an engine for them. He did so, but advanced to designing a complete aircraft, and as such he flew the *June Bug* on 21st June 1908. It was one of the first to use ailerons for lateral control, as opposed to the wing-warping as used by the Wrights.

His *Golden Flyer* set an air speed record of 52.6 mph (84.5 kph) in 1909.

Curtiss was the first to fly a practical water-borne machine, on 26th January 1911 at San Diego. Over the next few years he established a fine reputation for marine aircraft. They were widely used for patrol work in the First World War, including adapted versions built by Porte.

The crowning glory for Curtiss flying boats was the first Atlantic crossing by air, with a stop in the Azores, by **Albert Read** in the Curtiss NC-4 in 1919.

In the meantime Curtiss had produced the classic wartime American trainer, his JN-4 'Jenny.' Over 8,000 were built and almost all American pilots of the time would have handled one. They became popular 'barnstorming' machines in post-war years.

Curtiss, Glenn *(continued)*

The **Wright Brothers** accused Curtiss of patent infringements. He developed an antipathy towards them and sought to prove that **Samuel Langley's** *Aerodrome* could have beaten them into the air. He rebuilt it and flew it, but it later emerged that he had secretly improved it, invalidating his objective. The episode spoiled his otherwise fine reputation.

Curtiss died from appendicitis in 1930. Ironically his company later merged with that of his former foes to form Curtiss-Wright, but it failed to move into the jet age and faded in the 1940s, the engine side lingering on into the 1950s.

DAEDALUS – mythical Greek aviator. Daedalus and his son, Icarus, escaped from the Labyrinth prison on Crete. Ironically it was he who had designed the Labyrinth! The father survived, according to the legend, but the son was less fortunate.

D'AMECOURT, VISCOMTE (19th cent.) – experimented with helicopter models in the 1860s. The models achieved only limited success, but the Viscomte did make one contribution to history: he coined the word 'helicopter' (actually with the French spelling).

DASSAULT – **see Bloch, Marcel.**

DAVIES, STUART ('Cock') (1906-95) – Chief Designer, Avro 1947-1955. Davies survived the Tudor crash which killed **Roy Chadwick**, his predecessor. His best-known creation was the mighty delta-wing Vulcan bomber, which first flew in 1952. The aircraft was designed to meet a requirement for a long-range strategic bomber. In what may seem remarkable extravagance today, no less than three four-engine jet bombers were ordered into service, the Valiant, Vulcan and the Victor (the 'V-Bombers'), and a fourth type flew as prototypes, the Short Sperrin.

Davies said later that he adopted the delta layout almost by accident, by evolution after studying a number of wing shapes. Prudently a series of small research aeroplanes were flown, the Avro 707s, to test the delta behaviour before the Vulcan was completed. The delta wing proved a success, enabling the Vulcan to out-turn most fighters of its day, and giving ample space for fuel. Davies had designed one of the most impressive aeroplanes ever seen. It was even rolled spectacularly at low level at Farnborough.

Vulcans saw action just once, to bomb Port Stanley airfield in the Falklands in 1982. The return flight of 7,800 miles (12,500 km) made it the longest bombing raid made at the time (it has since been exceeded in the Gulf War). Despite the usual attempts at denigration, the first Vulcan raid scored a direct hit on the runway.

Davies handled the early design studies for the Avro (later Hawker Siddeley 748) twin-Dart powered airliner, which became one of the best-selling British commercial aircraft.

His nickname arose from his London (Cockney) origins whilst working among Mancunians.

DA, VINCI, LEONARDO (1452-1519) – sketched helicopters and ornithopters.The great artist and inventor drew helicopters using an Archimedes type screw for lift, and flapping-wing devices. For all his genius in other directions, none would have been practicable even if a suitable power source had existed.

DE HAVILLAND, SIR GEOFFREY (1882-1965) – pioneer pilot and aircraft builder. The first of the great line of de Havilland aircraft rose briefly into the air in December 1909 but crashed almost immediately. He had built both the airframe and engine but had no chance to practise piloting.

He flew his first successful machine on 10th October 1910. Lacking resources for development on his own, he joined the Farnborough team and then moved on to **Holt Thomas'** Airco company in 1914. There he designed, among others, the DH2 Scout and, most importantly, the DH4 bomber. This was first flown by de Havilland in August 1916 and proved to be one of the most successful bombers of the war. 1,450 were made in Britain, but as many as 4,844 were built in America, where it comprised the bulk of front-line aircraft in that war.

In 1920 the Airco company was sold and de Havilland set up his own business. He was keen to develop light aircraft and flew the first of his famous Moths in 1925. It was his interest in natural history and entomology in particular which led to the choice of names. He set up an engine division under **Major Halford** to produce the Gipsy series of engines.

Moths were used by numerous record-breakers, and were sold and built in many countries. 8,800 Tiger Moths alone were made.

Geoffrey de Havilland continued piloting himself during this period, one of the last of the founders of the British aircraft industry to do so. He was a skilful racing pilot and won the 1933 King's Cup. He retained his licence until 1953.

De Havilland, Sir Geoffrey *(continued)*

While preferring the peaceful civil aviation scene to the military market, he necessarily returned to the latter in the Second World War and under his guidance his company proposed the outstanding Mosquito. He met official resistance to the idea of an unarmed bomber, relying on speed and height for protection rather than guns, but it is to his credit that he persisted and was vindicated by the exceptionally low loss rates it sustained.

Another vital contribution to the war effort was his early appreciation of the merits of variable pitch propellers. He bought a licence from Hamilton Standard in the1930s, originally to give the necessary performance to the Comet racer which won the 1934 MacRobertson Race to Australia. A high proportion of wartime British combat aircraft were fitted with de Havilland propellers.

He was quick to see the possibilities of jets, and his Vampire fighter was one of the first in service, although just too late to take part in the war.

De Havilland soon resumed commercial aircraft work when peace returned with his Dove, and later Heron, light airliners. However, his company was now too big to return to the light aeroplanes for which it had once been renowned.

A bold step was his launch of the world's first jet airliner, the Comet. The airliner flew in 1949 and entered service in 1952, but all the hopes of his team were shattered by a series of accidents caused by fatigue failures of the pressure cabin.

De Havilland had built aircraft from frail wooden biplanes right up to the jet age. He died shortly after the last airliner to carry his name, the Trident had flown. From then on new aircraft would bear the Hawker Siddeley name, and ultimately that of British Aerospace, but now even his great factory at Hatfield has closed its doors.

There was sadness among de Havilland's many achievements, for two of his sons were killed in his company's aircraft. Peter lost his life in a collision in 1943, followed by Geoffrey in 1946 flying the tail-less DH108.

DERRY, JOHN (1922-52) – first to fly supersonically in Britain. Following wartime service in Fighter Command, Derry became a test pilot at Supermarine in 1947, moving to de Havilland later that year.
On 6th September 1948 he became the first pilot in Britain to exceed Mach 1, flying the tail-less DH108. How fortunate he was to survive the experience! The aircraft was highly unstable at speed and he only just kept control. Each of the three DH108s built killed a pilot.
A polished display pilot, he devised a manoeuvre which became known as the Derry Turn, in which the aircraft is rolled through 270 degrees to change direction.

His good fortune in surviving the dangerous DH108 testing was not to last. Whilst flying the DH110 (forerunner of the Sea Vixen naval fighter) at the 1952 Farnborough Show, the aircraft broke up. He was killed along with observer Tony Richards and 28 people on the ground.

DEWOITINE, EMILE (1892-?) – French aircraft designer. Dewoitine set up his company in 1920 and built a range of fighters and a successful airliner, the 22-seat D338 of 1936.

His D520 was the best French fighter of the Second World War, but upheavals caused by the nationalisation of the aircraft industry in the late 1930s had interfered with production and there were too few of them to slow the German advance.

Dewoitine emigrated to Argentina after the war and designed a jet fighter, the Pulqui 1. It flew but did not go into production and he faded into obscurity.

Bomber pilot Tom Dobney flew his Whitley into action over enemy territory before his senior officers discovered his secret - he had falsified his age and was just 15 years old.

DONALDSON, EDWARD – holder of post-war speed record. An outstanding aerobatic pilot, Donaldson flew in the Hendon display teams in 1935 and 1936.

He was station commander when the first RAF jet squadron was formed, in 1944. He was renowned as an excellent shot in combat.

On 7th September 1946 Donaldson set a new world speed record in a Meteor 4 of 616 mph. It was just below the 1,000 kph level which was much sought after at the time on the Continent.

He later became Air Correspondent for the *Daily Telegraph*.

Counting his Chickens?

A newspaper covering Donaldson's record captioned his picture with the word's "...Wing Commander Donaldson, seen here with his ground crew...." The picture actually showed a woman feeding her chickens.

DOOLITTLE, JAMES ('Jimmy') (1896-1993) – racing pilot and leader of first raid on Japan. In 1922, at the second attempt, Doolittle made the first flight across the USA in under a day, flying a DH4.

He was a determined racing pilot, winning the 1925 Schneider Trophy and numerous races in America. Some of his wins were in a lethal little racer, the GeeBee, looking like a huge engine with stub wings and unforgiving of mistakes.

After crashing an aircraft in poor visibility, he resolved to develop blind-flying and on 29th September 1929 made the first wholly 'blind' flight on instruments alone, with a second pilot to monitor the flight.

During the 1930s he worked for Shell and it was his influence which persuaded the company to produce 100 Octane fuel. It gave Allied combat aircraft vital extra performance in the war.

In April 1942 he led the famous carrier-borne bombing attack on Japan. 16 B-25B Mitchells were flown from the carrier *USS Hornet*. The bombers were not designed for carrier use and the take-off alone was a test of airmanship. There was no chance of returning and surviving crews went on to land in China. The raid was hailed as a triumph, although there was little real damage and there were casualties. Its value was principally psychological, on both sides, but in wartime that can be as important as physical damage.

Towards the end of the war he led the Eighth Air Force, doing much to improve tactics and reduce losses. Less popular with crews was his action in raising the number of missions per tour from 25 to 30 and again to 35. From 1956 to 1959 he was Chairman of NACA (National Advisory Committee for Aeronautics), the predecessor of NASA.

An outstanding aerobatic pilot, who may have made the first outside loop, Doolittle was less skilful in the bar, where he broke both ankles in a party game during a sales tour for Curtiss. He still gave his display the next day.

DORNIER, CLAUDIUS (1884-1969) – German aircraft builder. Dornier joined Zeppelin in 1910, and became Director of Design for the company's flying-boats in 1914.

He later formed his own company, building the first of the excellent Dornier Wal flying-boats in 1922. At first he assembled them in Italy due to restrictions on aircraft manufacture in Germany. The Wals performed many notable long-distance flights, and were employed on pioneering airmail services, using catapults on ships. The design was so sound that updated versions were flown in the 1980s.

His mighty Do.X flying-boat of 1929 stretched his ambition too far. Even with 12 engines it was underpowered and the three made were virtually unusable, although on one occasion an impressive load of 169 people, including nine stowaways, was carried.

Inevitably he was drawn into military work in the 1930s, and his best-known, or infamous, products of the war years were his Do.17 and Do.215 bombers, commonly known in Britain as the 'Flying Pencils' due to their slim fuselages.

In 1937 he had patented a fighter 'push-pull' layout, with one engine in the nose and another in the tail. The arrangement reduced drag and avoided asymmetric handling problems in the event of engine failure. It evolved into the remarkable Do.335 Pfeil, one of the fastest piston-engine aircraft ever made. To avoid the rear propeller shredding the pilot if he had to bale out, there was even an early ejection seat. It entered service but, as far as is known, never engaged in combat.

In post-war years he reformed his company, this time restarting in Spain for similar reasons to his previous exile. Eventually Dornier returned to Germany and built a line of light utility aircraft, airliners, and was a partner in the Alphajet trainer. His son followed him in the company.

Quicker by Sea

The Do.X flying-boat made one transatlantic crossing, but the time for the journey was hardly impressive - it took nine months!

DOUGLAS, DONALD (1892-1981) – founder of the Douglas aircraft company. American-born, though of Scots ancestry, Douglas started his long aviation career with **Glenn Martin** in 1916. Just after the war he formed his own company with a partner, David Davis (later the originator of an efficient wing design used on the Liberator, among other applications). His first aircraft was the Cloudster two-seat biplane, which luckily for him proved just right to meet a new US Navy requirement for a torpedo bomber. The great Douglas company was in business.

Douglas won renown in 1924 when three Douglas World Cruisers, out of an original four, completed the first flight around the world.

His real breakthrough came when the airline TWA asked Douglas for a three-engine airliner to compete with United's Boeing 247s. Douglas thought he could do the job better with a twin, and with his designer Arthur Raymond produced the 12-seat DC-1 (Douglas Commercial 1), which flew on 1st July 1933. The production version was the enlarged 14-seat DC-2, and it was so far in advance of other airliners that orders poured in from throughout the world.

American Airlines then asked Douglas for a widened version for long-distance sleeper services. Donald Douglas at first saw little market for it, but president Cyrus Smith of American persisted. How glad Douglas must later have been been that he did! The new airliner flew on 17th December 1935 and became one of the most important aircraft in aviation history, the DC-3, or Dakota as it was known in Britain.

Douglas now became the principal airliner supplier to the world and kept that leadership until the end of the piston era. As his company grew he also became a leading builder of fine military aircraft, including well-known types like the Boston wartime bomber, Dauntless naval bomber, A-1 Skyraider, and the A-4 Skyhawk jet strike fighter.

Douglas entered the jet airliner age with the DC-8, but he was more cautious than Boeing who were ahead with their famous 707 and took the greater part of the market. With the smaller DC-9, Douglas had a winner which became one of the most successful of all jet airliners, but once again he was cautious and at first did not want to build it. Successful as it was, the development costs of new airliners were straining the company and it merged with McDonnell in 1967.

The Immortal DC-3

The Douglas DC-3 is a true classic of aviation. By 1939 it had swept the board with airlines all over the world and was handling the bulk of all passenger traffic. On the outbreak of war, thousands were built as transports and used for every conceivable task, including dropping paratroops and glider towing.

Including production in Russia and Japan, Douglas believe as many as 20,000 may have been made.

In early post-war years the DC-3 again dominated the airways of the world, until larger and faster airliners displaced it on to secondary routes. Many a manufacturer tried to build a 'DC-3 replacement' , but many were outlived by the DC-3 itself. 60 years after it first flew, it is believed over 1,000 are still flying.

Even in the 1990s some operators have been fitting new turbine engines to their DC-3s, clearly expecting to fly them for many more years to come. It is a remarkable tribute to Donald Douglas and his team, and that was the aeroplane he didn't want to build!

DOUHET, GUILIO (1869-1930) – advocate of strategic bombing. In 1911 Douhet flew in a bombing raid on Turkish forces, possibly the first ever aerial bombing mission.

In 1921, by then a general in the Italian army, he wrote *The Command of the Air,* in which he declared that wars could be won by bombing alone, and that no effective defence against a bomber was possible.

His views were influential in many countries, but the development of radar rendered his theories obsolete.

Nevertheless, strategic bombing was widely, but controversially, used in the Second World War. In a sense, the atomic bomb gave his theory new validity.

DOWDING, SIR HUGH (Lord Dowding) (1882-1970) – Head of Fighter Command 1936-40. Originally a soldier, Dowding qualified as a pilot in 1914 at the then high age of 32. He flew operationally and held staff posts in the First World War.

In 1930 he became the Air Member for Supply and in 1936 came the appointment as Head of Fighter Command.

Hugh Dowding – masterful tactician of the Battle of Britain

He saw the need for the new eight-gun fighters being developed by industry and urged the large-scale production which was to prove so vital. He faced heavy opposition, for surprising as it may seem today, there were many in the RAF who believed that biplanes would hold their own against monoplanes thanks to better manoeuvrability.

After allegedly early scepticism, he also rightly saw the importance of radar and pressed for the early-warning chains to be put into service. Without them, it is difficult to see how the Battle of Britain could have been won.

In the 1940 Battle itself, his skilful handling of resources, including resisting the clamour to waste fighters and crews on a futile defence of France, has been acknowledged by all historians, although it was a 'close-run thing'. He was less successful in stopping a destructive feud about tactics between his two most crucial group commanders, **Park** and **Leigh-Mallory.**

Undoubtedly he could be abrasive, but the abrupt removal in November 1940 of Dowding and **Park** caused deep resentment among those who had served with these two men. He was in fact overdue for relief from his post and he had been under great strain, but the manner of his replacement provoked offence.

His vital contribution to saving Britain was later acknowledged when 'Stuffy' Dowding was made a baron in 1943.

DUKE, NEVILLE (1922-) – test pilot and world speed record holder. A wartime fighter pilot, Duke achieved at least 28 victories, mainly in the Middle East. He was shot down three times, once being lucky to survive when he landed in a lake and was nearly drowned by his parachute.

He became a test pilot after the war, joining Hawkers in 1948. On 7th September 1953 he set a world speed record of 727.6 mph (1170 kph) in the sole Hunter F3, painted bright red and specially fitted with an afterburner on the Rolls-Royce Avon engine. He became a popular household name after this exploit – a number of test pilots were national celebrities at the time in a way almost unknown today.

He retired from test flying of jets in 1956 due to a back injury sustained in a Hunter accident, although he continued flying business aircraft and some freelance testing of light aeroplanes.

DUNNE, JOHN (1875-1949) – first to fly a swept-wing, tail-less aeroplane. Service as a soldier in the Boer War, during which he was wounded, convinced Dunne of the potential of aerial reconnaissance. He believed that an aeroplane should be inherently stable, and his designs were influenced by the behaviour of the seed-bearing leaf of the African Zanonia tree. The results were a series of gliders and powered machines with no tails but markedly swept wings.

He started his work at Farnborough, but no military funds were forthcoming and he moved his experiments to the Blair Atholl estate in Perthshire. In 1907 he flew gliders, followed by a powered aircraft, his D5, built for him by Shorts. He worked hard to persuade sceptical senior army officers of the value of aeroplanes in war.

The stability of his aircraft was such that he demonstrated it by flying with his hands above his head. He analysed the results of his flights carefully and even wrote notes in the air, in a sense becoming the first test pilot.

Licences were granted to built Dunne designs in France and America. In 1923 failing health forced him to end his experiments. Although his quest for stability proved a fallacy, his designs and analytical approach to testing were in many ways ahead of his time.

The Mystery of the Moors

The Blair Atholl estate had one great advantage for Dunne's testing: it was secluded. Soon rumours started about secret flying machines being flown there, and the press made much of 'the mystery of the moors'. It is said that German and other spies were seen, while gillies and local boys were paid to guide undesirable visitors in the wrong direction.

DUNNING, EDWIN (?-1917) – first to land on a moving ship. A flying deck had been built on the foredeck of a cruiser, *HMS Furious.* On 2nd August 1917 Dunning flew a Sopwith Pup round the superstructure and landed on the deck. The ship was steaming at 26 knots into a headwind.

Flying around the ship's structure and through the turbulence it caused was clearly hazardous, and sadly this became all too apparent when he tried to repeat the feat a few days later. The Pup went over the side, despite desperate efforts by officers on deck to hold it, and courageous Dunning was killed.

His death was not entirely in vain, since after this accident the superstructure was removed and the aircraft carrier was born. *Furious* went on to a long and distinguished career, including plenty of action in World War Two, and was not scrapped until 1949.

EARHART, AMELIA (1898-1937) – female record-breaker. The first woman to cross the Atlantic, albeit as a passenger, was Amelia Earhart in 1928, in a Fokker VII flown by Wilmer Stultz.

She was not the sort of person who was happy to accept a 'first' as a passenger, and she was determined to make a crossing as a pilot. She realised her ambition in 1932, with a fine solo flight in a Lockheed Vega. This was the first solo Atlantic flight by a woman.

Earhart now made a series of record-breaking flights, including two coast-to-coast crossings of America and a flight from Hawaii to San Francisco.

The culmination of her long-distance records was to have been a round-the-world flight in 1937. On her second attempt, she set out in a Lockheed Electra with navigator Fred Noonan. One of the most critical stages was a stretch over the Pacific to the small Howland Island. They never arrived. Many searches were made, and are continued spasmodically to this day, but no evidence for their fate has been found.

Despite sensational theories about being imprisoned by Japanese forces, the most likely explanation is the mundane one that they missed the island. Noonan may not have been the ideal choice of companion, for he had a drink problem.

ECKENER, DR HUGO (1868-1937) – leading exponent of airship travel. A designer and pilot working for the Zeppelin company in the First World War, Eckener became managing director after the death of its founder.

Germany was forbidden to build airships between 1918 and 1925. When the ban was lifted, Eckener was the driving force in restarting an airship programme and passenger services. Notably he raised money by public subscription to build the *Graf Zeppelin* in 1928.

In 1929 he organised a flight round the world to show the *Graf Zeppelin's* capabilities. It took 12$^{1}/_{2}$ days, and included a couple of 'close calls'. Thereafter he ran a scheduled service to South America with the airship. It was technically successful, but the economics with 40 crew to look after 20 passengers must have been questionable.

Under his direction the larger *Hindenburg* entered service on the North Atlantic. Despite Eckener's lobbying, America refused to export helium due to Hitler's bellicose behaviour, and both airships used hydrogen. The end came on 6th May 1937 when the *Hindenburg* exploded at Lakehurst. Whether it was caused by a gas leak ignited by static electricity or a cable snapping, or by sabotage, was never proven. It was truly remarkable that 62 of the 97 aboard survived.

After the *Hindenburg* disaster the *Graf Zeppelin* was scrapped and Eckener's dreams of airships linking the continents were over.

EDWARDS, SIR GEORGE (1908-) – designer of the Viscount. Edwards joined Vickers in 1935 and became chief designer in 1945. The company had in hand a modest post-war airliner, the Viking, but he was charged with a more ambitious successor.

Because of Britain's priorities in building combat aircraft during the war, America had forged ahead of the rest of the world in airliner design. Edwards took the bold step of selecting turbine power in a bid to leap ahead. It was a risk, since turbine power was untried in passenger service and there were worries about the durability of the engines and airline reaction to the unknown. He chose the Rolls-Royce Dart turboprop, a decision which must have given him loss of sleep when the engine first ran and was both grossly overweight and low on power!

His faith was justified and the Viscount was a winner. The prototype flew the world's first turbine-powered service on 29th July 1950, although full operations had to wait until 1953. 444 were sold, including substantial orders in Canada and the United States, and a batch for China.

His later Vanguard and VC10 airliners sold in small numbers only, but his BAC One-Eleven fared better, following the Viscount with useful sales in America. Nevertheless it did not reach his original hopes, due to a series of accidents during test flying and an inability to 'stretch' the airframe and its Rolls-Royce Spey engine to match its competitors.

Edwards, by then Sir George, became managing director of the civil division of BAC, and handled much of the planning for Concorde. He retired in 1975.

At a time when Concorde was under criticism, Sir George claimed he had a supporter on high, when he quoted from the Prayer Book ".....author of peace and lover of Concord(e)"

EILMER (12th cent.) – flying monk of Malmesbury. The monk, presumably of great faith or courage, launched himself from the top of Malmesbury Abbey, possibly in 1110. He reputedly flew a furlong and survived, but 'maiming all his limbs' on landing.

Creditably, in best test pilot and designer practice, he tried to diagnose the problem and concluded it was 'for want of a tail'. Prudently, he does not seem to have repeated the experiment.

ELY, EUGENE (1886-1911) – first to take-off from and land on a ship. The take-off was from a platform set up on the bows of *USS Birmingham* on 14th November 1910. The ship was stationary.

He only just succeeded, for his Curtiss biplane touched the water before flying away.

The landing took place on 18th January 1911 on a platform over the stern of the *USS Pennsylvania*, also whilst the vessel was stationary.

Ely was killed later that year, performing stunts at a display.

ELLEHAMMER, JACOB (1871-1946) – the Danish pioneer. A brief 'hop' by Ellehammer on 12th September 1906 was at one time claimed as the first aeroplane flight in Europe, but few historians count it as a proper flight.

The aircraft was a biplane powered by an engine of his own design. The trials were conducted on a circular track, with the aircraft tethered to a central pole.

His original method of take-off was seriously considered again during the Second World War, for use where long runways would have been difficult to build.

Ellehammer is credited with the first flight in Germany, in a later triplane, on 28th June 1908.

ESMONDE, EUGENE (1909-42) – led Swordfish attacks on German warships. Irish-born, Esmonde flew pre-war airline services on Short 'C' Class flying-boats with Imperial Airways.

On the outbreak of war he joined the Royal Navy. He led an attack on the battleship *Bismarck* by Fairey Swordfish biplanes. For all their bravery, little damage was done, but later other Swordfish pilots scored a torpedo strike which disabled *Bismarck's* rudder and led to her sinking by surface ships.

On 4th February 1942 he led the almost suicidal attack on the German cruisers *Gneisenau* and *Scharnhorst* in the English Channel. The promised fighter escort failed to appear, and although Esmonde would have been justified in aborting the mission he pressed on and every Swordfish was lost. He was awarded a posthumous VC.

The Courage of the Swordfish Crews

The bravery of the men who flew those slow, open-cockpit biplanes cannot be overstated. Despite their obsolete equipment they caused immense damage to enemy shipping. It has often been claimed that German gunners could not easily aim on them because they were so slow!

ESNAULT-PELTERIE, ROBERT (1881-1957) – French pioneer. One of the most visionary of early aviators, Esnault-Pelterie started experiments in 1904 with a glider based on the little known about the **Wrights**' work. He fitted ailerons on this glider, perhaps the earliest in the world.

The results were disappointing, although he made one advance in trying to launch behind a car, later to become, with the similar use of winches, a standard method of glider launching.

He flew a powered monoplane in 1907, the R.E.P.1, named after his initials. With its internally braced wings (i.e. without struts or wires), and fully enclosed steel-tube fuselage, it looked years ahead of its time. His engine was a fan-shaped seven-cylinder unit, and here again he led the way for the air-cooled radials to come. He was also the first to recognise how an odd number of cylinders in such engines would offer smoother running and higher efficiency.

The R.E.P. was underpowered, but later versions with more powerful engines performed well. Among the many features he pioneered were sprung undercarriages, the modern layout of stick and rudder controls, and seat belts. The last-named may have saved his life in an accident in 1908, but he was nevertheless severely injured and gave up flying himself.

For all their advanced features, or perhaps because they increased the costs, his aeroplanes never sold well and his works was sold to **Farman** in 1913.

Even while he was experimenting with these early aircraft, he was predicting spaceflight and writing papers on the subject. It was a sad coincidence that he died in the very year that the first artificial satellite was launched into orbit.

FAIREY, SIR RICHARD (1887-1956) – British aircraft constructor. After working with **John Dunne** in 1910, and for a while with the **Short Brothers**, Fairey set up his own company in 1915. Most of his early work was building Sopwith and Short types, but before long he started on his own designs. He is widely credited with having devised the first wing flap during the war.

Soon his designs were well regarded, and his Fairey III series was widely used by the British services and overseas. In 1925 he produced his Fox light bomber, faster than any fighter at the time, with a single Curtiss D12 engine. That engine was advanced for its day and Fairey wanted to build it under licence. Air Ministry permission was refused as they did not wish to introduce more engine companies in Britain. Fairey was disappointed that sales of the Fox were limited to just one squadron for the RAF and a Belgian order.

In 1927 he set up a small aerodrome west of London. It is rather bigger now – as Heathrow.

He had a success when a specially built Fairey Long-Range Monoplane set a world record on 8th February 1933 by flying 5,309 miles (8,544 km) non-stop from Lincolnshire to Walvis Bay.

His next types had mixed fortunes. The 1936 Fairey Battle light bomber looked modern for its day, but four years later proved too slow and lightly armed. Its crews suffered terribly. However the far more antiquated-looking Swordfish biplane achieved some spectacular wartime successes. Remarkably enough, it was one of few pre-war types still in production at the end of the war.

Fairey retired in 1945. His aircraft business was sold to Westland in 1960, but other divisions making hydraulic and marine equipment continued trading under the Fairey name.

'Roly' Falk, test pilot for Avro, and 'Tex' Johnson of Boeing, both created sensations by rolling four-engine jet aircraft at air displays, Falk in the Vulcan and Johnson in the Boeing 707. Incidentally Falk flew a '707' too, the Avro 707 research delta.

FARMAN, HENRY (1874-1958) and **MAURICE** (1877-1964) – pioneers of French aviation. There were actually three brothers, Henry, Maurice and Dick, but it was the first two who became well-known in early French flying circles. At the risk of causing upset in France, it must also be recorded that they were English-born, but they settled in France. Henry was better known as Henri, and became a French citizen in 1937.

Henry bought a Voisin biplane in 1907, but modified it considerably. He flew the first recorded European one kilometre circuit on 13th January 1908. It could not have been easy, for there was no means of lateral control on this aeroplane.

Henry and Maurice set up separate companies in 1909, which merged again in 1915. Farmans were flown by many pre-1914 pioneers, including some in Britain, and they formed the basis of early Bristol designs. Farmans were also built in large numbers during the First World War, but their design stagnated somewhat due to French government insistence on 'pusher' types.

After the war the brothers turned to airliners and bombers. The Goliath was a mainstay of French airline services for some years. Passengers enjoyed the unusual experience of front-facing windows. The company was nationalised in 1937.

Maurice Farman was still flying until well into his 80s. As he had never applied for a licence, in later years he had to be accompanied by a qualified pilot.

FEDDEN, SIR ROY (1885-1973) – aero-engine designer. In the years before 1914, Fedden learned his engineering in the car industry. He designed a car, the Brazil Squire, which sold well.

During the war, Brazil under Fedden built Rolls-Royce engines under licence. It is a tribute to his ability that he was the only man so entrusted by **Henry Royce**.

Under the terms of his licence, Fedden could not design liquid-cooled engines of his own, so he turned to the air-cooled radial. After changes in company ownership, Brazil became the Engine Division of the Bristol Aeroplane Company, with Fedden as its head.

He received a commission of a few pounds on every engine sold. Later this clause in his contract caused jealousy, when wartime production churned out engines by the tens of thousand.

His Jupiter engine became world renowned, was chosen by aircraft builders in many countries, and built overseas by 17 licensees. One airline bought his later Pegasus for its Douglas DC-2s, but unfortunately no further airlines followed suit.

In the early 1930s he became attracted to the sleeve valve. The idea had been used in a few cars, but Fedden was the first to try it on aero-engines. After years wrestling with technical problems, he succeeded and over 100,000 Hercules and Centaurus sleeve valve engines were made in the war. Famous aircraft like most Halifaxes, Beaufighters, and Wellingtons relied on the Hercules. They were also used in post-war airliners, some serving into the 1990s, and reaching record piston engine overhaul lives of 3,000 hours.

A strong personality, he worked himself and others to their limits. In 1942 he received a knighthood, then just days later the Bristol Aeroplane Company dismissed him. He was not even allowed to remove personal items from his office. His abrupt departure slowed development of the vital Bristol engines at a crucial time, and delayed the company in making the move to jets.

After the war he set up a company which designed cars, light aircraft engines and a turboprop. By tackling too many projects at once without an income, it ran out of money. He then worked for NATO and handled consultancy work.

Air or Liquid

Fedden was one of the great exponents of the air-cooled radial engine. Passionate arguments raged between supporters of liquid and air cooling, and still do on occasion. At the risk of causing umbrage even now, the air-cooled unit with its lower build and overhaul cost had the advantage in transports, while the liquid-cooled was regarded as better for combat aircraft because of its lower frontal area.

However, this advantage was partly offset by the drag of the radiator, which was also vulnerable in battle. On balance the advantage lay with the air-cooled radial, just, but nothing can detract from the record of the liquid-cooled Merlin.

One of Fedden's quirks, of which today's legislators would not approve, was his refusal to employ any women in his works. Typists and secretaries were all men.

FERBER, FERDINAND (1862-1909) – pioneer French aviator. Inspired by Lilienthal, Ferber began gliding trials as early as 1901, without real success. On hearing about the Wright Brothers' achievements, he tried to interest the French government in military uses of the aeroplane.

Ferber flew a design of his own in July 1908, but was killed in an accident the following year. His contribution to aviation history was not so much in his own trials, which were not very successful, but in making the work of the Wrights and Octave Chanute known in France and so inspiring the early French aviators in their efforts.

FIESELER, GERHARD (1886-1987) – German aircraft constructor. A World War One fighter pilot, Fieseler won renown in post-war years as an outstanding aerobatic pilot.

In 1930 he entered the field of manufacturing by buying a sailplane company. In the early 1930s he designed, as a private venture, a light utility aircraft with large flaps and leading-edge slats to give outstanding take-off and slow flying capability. First flown in 1936, the Fi 156 Storch became easily the most famous aeroplane associated with his name. It was widely used by German units during the war, and two missions in particular are regarded as flying epics. The first was the rescue of Mussolini from a supposedly impregnable mountain fortress, where he was held prisoner, by a German commando unit landed by Storch. The other episode was a daring night flight into Berlin in the last hours of the war by the renowned woman pilot **Hanna Reitsch**.

A most unwelcome visitor to Britain was another of Fieseler's products, the Fi 103, better known as the V1 pilotless bomb.

FLEET, REUBEN (1887-1975) – founder of the Consolidated company. Fleet organised American pilot training during the First World War, an experience which inspired him to build better trainers of his own design. To this end he formed the Consolidated Aircraft Corporation in 1923. By a policy of mergers, he developed it into one of the 'giants' of the American aircraft industry.

Under his management he saw the PBY Catalina flying-boat and B-24 Liberator programmes started, but he left the industry before they reached their peaks. Consolidated later became Convair and later still General Dynamics.

FOKKER, ANTHONY (1890-1939) – Dutch aircraft builder. Originally from the Dutch East Indies, Fokker started his long career in aircraft manufacture in Germany. He flew his first design, a monoplane, in 1911. On the outbreak of war in 1914, he was apparently not too fussy about which side to serve. He claimed he offered his output first to the Allies, then to the Germans.

Fokker gained notoriety amongst Allied aircrew with his Eindekker ('Monoplane') in 1915. He was shown the rather crude deflector plates, which had enabled a machine-gun to fire through the propeller disc, salvaged from **Roland Garros'** shot down aircraft. He improved upon the idea with his interrupter gear, which synchronised the firing of the gun with the position of the propeller blades. According to his own account, he was ordered to prove the device by shooting down an Allied aircraft, an astounding order to a nominally neutral civilian.

He stated that he positioned himself ready to fire then balked at killing the occupants. Whatever the facts, his monoplanes so fitted proved lethal and the period became known as the 'Fokker Scourge'.

Aided by his talented designer, Reinhold Platz, Fokker followed with a series of biplanes and triplanes. One of the latter went down in history as the last mount of **von Richthofen**, but in fact the triplanes were made in fairly small numbers and their performance was not outstanding. Much more effective was his Fokker D VII biplane of 1917, one of the best fighters of its time.

At the end of the war, Fokker was ordered to destroy his factory and the partly built aircraft within. He secretly moved them at night to Holland and restarted production there.

Post-war Fokker high-wing monoplanes became widely used and achieved many notable 'firsts'. Amongst these were the first non-stop flight across the USA in 1923, **Byrd's** North Pole flight, and **Kingsford-Smith's** 1928 crossing of the Pacific. By the mid-1930s Fokker realised his wooden airliners were becoming obsolete and he took out a licence to build the Douglas DC-2, although production in Holland never started. He died just before the outbreak of war.

His company restarted in post-war years and built some excellent airliners, but finally proved too small for the increasingly complex airliner business and its long history ended in 1996.

FOLLAND, HENRY (1889-1954) – aircraft designer. Folland first made his name as a designer at the Royal Aircraft Factory with his excellent SE5a fighter. Fast and manoeuvrable, it was one of the best of the war.

In the 1920s and 1930s he designed the range of Gloster biplane fighters and racers, culminating in the Gladiator, the last RAF biplane fighter. Some saw combat in the Second World War and scored a surprising number of victories. Three Sea Gladiators, named Faith, Hope, and Charity, formed the sole air defence of Malta for a couple of weeks.

Folland formed his own company in 1937. It mainly built other companies' designs and components, but he completed a small number of his own prototypes.

Folland, Henry *(continued)*

He retired in 1951 but it is ironic that the best-known aircraft to carry the Folland name was designed by his successor, **Teddy Petter**. This was the Gnat lightweight fighter and trainer, successful in action in India and famed as the mount of the Red Arrows for many years.

> ***Henry Folland's talents seemed to desert him in some of his designs of the late 1930s. One was so disliked by pilots that they called it the 'Folland Frightful', with some reason as it broke up, injuring the pilot.***

FONCKE, RENÉ (1894-1953) – top-scoring French pilot of World War One. Often classed as the greatest fighter pilot of the war, Foncke is credited with between 75 and 120 victories. The true figure is uncertain as he sometimes credited novice pilots with his own 'kills'. On two occasions he scored six in a day. Such figures imply just how short a life expectancy lesser or unluckier pilots faced.

In 1926 Foncke attempted an Atlantic flight in a Sikorski S35. The heavily-laden aircraft failed to become airborne, ground-looped and caught fire. He survived but two crewmen did not.

He helped in a belated attempt to expand the French Air Force before the Second World War, retiring just before hostilities started. Sadly this great Frenchman was unjustly accused of collaborating with the Vichy regime.

FOZARD, PROFESSOR JOHN (1928-) – introduced the 'ski-jump' launch technique. A designer for Hawker since 1950, and later Hawker Siddeley, Fozard took over development of the Harrier after **Sydney Camm's** retirement.

A significant advance was his implementation of the 'ski-jump' ramps on the bows of ships launching Harriers. It was based on an original idea by Lt-Cdr D Taylor. The ramp allowed the Harrier to take off at much higher weights, giving more range or heavier loads.

Such ramps were fitted on the bows of *HMS Hermes* and *Illustrious* when they were deployed to the Falkland Islands in 1982. The extra performance was invaluable in that campaign.

FRISE, LESLIE (1897-1979) – designer with the Bristol Aeroplane Company. Frise joined Bristols during the First World War and worked with **Frank Barnwell,** succeeding him upon the latter's death in 1938.

He invented the Frise aileron in 1932. The 'down' aileron tends to create more drag on the outside of a turn than the other. Frise hinged his aileron so that when an aileron was raised, part of the surface projected below the wing, compensating for the difference in drag between the two wings. It became widely used, until **Arthur Hagg** devised differential ailerons.

Frise was responsible for the fast and heavily armed Beaufighter, and later the Bristol 170 Freighter. The sturdy high-wing transport was the mainstay of cross-channel car ferry services in the 1950s. A Channel crossing took 20 minutes then – beat that Eurotunnel, 40 years later!

In 1948 he moved to Hunting Percival and designed the Provost and its more important development the Jet Provost, a major element of RAF training for nearly 40 years.

FUCHIDA, MITSUO – led the Japanese attack on Pearl Harbour. The attack comprised a force of 183 aircraft. Whatever the ethics of what Americans called 'a day of infamy', the attack itself was carefully planned and executed.

Several capital ships were sunk, with heavy loss of life, but in one respect it failed. The vital American aircraft carriers were at sea and were untouched. The Japanese were to pay heavily for their escape.

GAGARIN, YURI (1934-68) – first man in space. The first manned spaceflight took place on 12th April 1961. Gagarin was aboard a Vostok spacecraft for a single orbit of 1 hour, 48 minutes. The wording 'was aboard' is used advisedly, for he did no actual piloting, unlike the first American astronauts.

Not until 1978 was it disclosed that he left the spacecraft at 22,000 ft (6,700 m) on the descent and landed by parachute.

The flight shook the West, and the Soviet Union exploited their success by parading the good-looking Gagarin around the world. He was killed flying a jet fighter in March 1968.

GALLAND, ADOLF ('Dolpho') (1912-96) – leading Luftwaffe fighter pilot. Like most German fighter pilots of his time, Galland started his flying career on gliders due to restrictions on powered flying in the country. He learned to fly in 1927 and later became an airline pilot.

He flew with the German Legion Kondor in the Spanish civil war, that training ground which taught so many lessons in tactics to individual pilots and the Luftwaffe as a whole.

He led a squadron in the Battle of Britain and showed himself to be one of the ablest of German fighter pilots. Over his combat career he is credited with at least 52 victories, and over 100 according to some sources.

At the end of 1941 Galland became General of Fighters, a post which included recommending which new types should be selected. He was keen on the Messerschmitt 262 jet, but took high personal risks in pressing for it to be used as an interceptor against Hitler's policy of wanting it as a bomber. Unusually, he survived disagreeing with the Führer and was appointed to lead an élite unit to fly these advanced fighters.

In post-war years he helped in establishing the new German Air Force, and became a popular figure in gatherings of former combats pilots from both sides.

GARNERIN, ANDRÉ JACQUES (1769-1823) – first to descend by parachute from a balloon. Garnerin was an early balloonist, having made his first ascent in 1787. After the French Revolution he served as a soldier and became a prisoner of war in Hungary. He had heard of a parachute drop from a tower by one Sebastian de Normand and dreamed of using such a means to escape.

After his release, he resolved to make a parachute descent from a balloon. On 22nd October 1797 he ascended 3,000 ft (920 m) attached below a balloon. On release the parachute opened and he descended safely, although swinging so violently that he felt sick – the need to make a canopy pervious to air was realised later. The founder of the Musée de l'Air in Paris described the event as 'one of the great acts of heroism in human history', and who would disagree?

In 1802 he demonstrated his parachute in London, reportedly attracting the largest crowd the city had ever seen up to that time.

Garnerin visited many countries giving both ballooning and parachuting demonstrations. He often made night ascents, sometimes further enlivening procedures by releasing fireworks from the air – a practice which must have been almost as hazardous from a hydrogen balloon as entrusting himself to his parachutes!

André Garnerin fell from favour and caused an international incident by a freak mischance. At Napoleon's coronation in 1804 he released a huge unmanned gas balloon, elaborately decorated and surmounted by a large gilded crown.

It could have landed anywhere in Europe, but by some extraordinary chance it landed in Rome.

It touched down in the Vatican, depositing the crown on Nero's tomb, before rising again and ending in a lake.

The Italians were much aggrieved, as was Napoleon for a time, although Garnerin was later restored to favour.

GARROS, ROLAND (1888-1918) – pioneer French aviator and fighter pilot. Unlike most of the early French pilots, whose backgrounds tended to be fields like cycle or motorbike racing, Garros was a concert pianist. He learned to fly from **Santos-Dumont** in 1910 and soon established himself in racing and record-breaking.

In 1913 he made the first flight across the Mediterranean. He calculated his fuel would last for eight hours – he landed in Tunisia after 7 hours 53 minutes! Safety margins in 1913 were thin.

During the war he sought ways of mounting forward-facing machine-guns on an aircraft without wrecking the propeller. With aircraft builder Raymond Saulnier he fitted metal deflector plates to the propeller blades, designed to avoid the risk of bullets ricocheting back into the structure. It was first used in 1915, taking unwary German pilots by surprise. Garros had in effect invented the fighter.

His idea soon proved its worth. His combat record is believed to have started the use of the term 'ace' for a pilot with five or more victories. Not long after, he was forced down behind enemy lines by engine trouble. Despite his efforts to destroy the machine, the Germans found the deflector plates which inspired **Anthony Fokker** in devising his improved interrupter gear.

Garros was captured but later escaped. He was killed in October 1918, a sad loss when peace was so near.

GIBSON, GUY (1918-44) – leader of the 'Dambusters'. A pre-war RAF pilot, who ironically was originally turned down by the service, Gibson had already served three 'tours' of operations, two on bombers and one on nightfighters, when he was chosen to lead an élite unit for one mission. The squadron was formed as number 617 and he was able to pick the best crews he could find.

After weeks of low-level training in their Lancasters, the target was revealed as a group of dams in southern Germany. They were to be destroyed by **Barnes Wallis**' 'bouncing bomb', which had to be released at just 60 ft (18 m), a most demanding requirement at night.

Two of the dams, the Mohne and Eder, were breached and considerable damage was done to German industry. After dropping his own bomb, Gibson flew alongside some of the later crews to distract the defending gunners. On the debit side, 8 of the 19 aircraft were lost, an attrition rate that would be totally unacceptable on regular operations. Gibson was awarded a VC.

Gibson had done far more than his quota of missions but insisted on flying a few more. He was killed whilst flying a Mosquito on 19th September 1944. It was his 177th operation.

The Deluge of Debate

In recent years historians have questioned the military value of the dams raid. It is true the most important dam, the Sorpe, was not breached – its construction made it less susceptible to the bouncing bomb. The results were also achieved at the cost of heavy RAF and German casualties. The value of the raid was largely psychological – it was a tonic to the British people and showed what precision bombing could do. In wartime such considerations are important and many still feel the raid to have been of military value even though industrial damage was repaired fairly quickly.

The term 'Dambusters' seems to have been coined by Avro, makers of the Lancaster, at a dinner to honour the crews.

GIFFARD, HENRI (1825-82) – builder of the first airship. The dirigible was fitted with a 3 hp steam engine which gave it a speed of 6 mph (9.5 kph). Needless to say it could be flown only in the lightest of winds, but on 24th September 1852 he flew 17 miles (27 km) from Paris to Trappes. This made him the first man to fly with some control over where he was going.

Later he designed a monumental 7 million cubic feet airship, about the size of the *Hindenburg*, the largest ever made. It would have been powered by a 30 ton steam engine. Potential backers took fright at the enormous cost and risks, probably with good reason.

GLENN, JOHN (1921-) – first American to orbit the Earth. As a pilot with the US Marine Corps, Glenn flew 59 combat missions in World War II and a further 63 in Korea.

In 1957 he established a single-seat transcontinental record of 3 hours, 23 minutes from Los Angeles to New York, flying an F-8 Crusader.

His spaceflight was on 20th February 1962 in the Mercury capsule *Friendship 7*. He completed three orbits.

Later he became a Senator and made a brief bid for the Presidency.

GODDARD, ROBERT (1882-1945) – liquid-fuel rocket pioneer. Solid fuel rockets had been used since the 13th century, probably originating in China. They suffered the drawback that there was no control over the power once the rocket had been fired. It was Goddard who sought to overcome this problem by using liquid fuels.

While otherwise inactive due to childhood illness, Goddard had thought about the possibilities of spaceflight. In 1919 he published his paper *A Method of Reaching Extreme Altitude*, detailing the possibilities of liquid-fuel rocketry. In 1926 he was rewarded by the first flight of a rocket so powered. Later flights reached heights of several thousand feet, laying the foundations for the great space launchers to follow, all of which have used the liquid fuels he rightly advocated.

GOERING, HERMANN (1893-1946) – Luftwaffe commander. Originally a soldier, Goering became a fighter pilot during the First World War. He was wounded in 1917, but returned to command **von Richthofen's** old 'flying circus' in 1918. However distasteful his later life may have been, he was a courageous combat pilot, even if his frequent mishaps may have pointed to something lacking in his airmanship.

Goering joined the Nazi party in 1922 and was wounded again in the 1923 'putsch'. He became commander of the new Luftwaffe and wielded power second only to Hitler. Among his more unsavoury actions were the founding of the Gestapo and setting up the infamous concentration camps. He enjoyed an extravagant lifestyle with personal palaces, commandeered art collections, and his sumptuous private train. An obvious effect of his mode of living was that he became grossly overweight.

It was of the greatest fortune to the world that Goering's leadership of the Battle of Britain was flawed. He committed tactical blunders like ordering fighters to fly in close escort to his bombers instead of allowing them to use their height and speed, and switching bomber attacks from airfields to cities gave Fighter Command relief when its resources were nearly exhausted. Many strategists believe that the Luftwaffe would have won had it been as well led as the RAF.

He was committed for trial at Nuremberg but had managed to conceal some poison capsules and used them to take his own life.

GOUGE, SIR ARTHUR (1890-1962) – aircraft designer. Gouge is principally remembered for his fine flying-boats, first with Shorts and later at Saro. It is easy to believe that British airliner design was far behind that of American companies in the 1930s, and while that may be true of landplanes, Gouge's Short C Class 'Empire' flying-boats were amongst the most advanced aircraft in the world at the time. First flown in 1936, they offered a leisurely, luxurious form of air travel never seen before or since.

As war loomed, he turned his peaceful C Class 'boats into the heavily-armed Sunderland, the guardian of the convoys and respected by German aircrews who called it the 'flying porcupine'.

More aggressive still was the first of the British wartime heavy bombers, his four-engine Stirling. Had he been given a free hand in design, no doubt it would have been a better bomber than it was, but his hands were tied in two respects. Firstly, it was compromised by a requirement that it double up as a transport, and secondly that the wingspan was supposedly limited to fit within standard RAF hangars of the day.

The limited span restricted the height Stirlings could reach and made them vulnerable. Recently doubt has been cast on this origin of its short span.

Gouge left Shorts in 1943, following nationalisation of the company, and joined Saunders-Roe. He continued working on flying-boats, notably the mighty ten-engine Princess. It was well designed and in some respects like its use of powered controls led the world, but by then most cities had land airports and airlines were cool towards new flying-boats.

One of the three built flew in 1952 but thereafter they lay like beached whales on their slipways for 15 years.

Still clinging to marine aircraft, he designed a unique jet flying-boat, the Saro SR.A/1. It flew well, but the drag of its hull penalised it compared with land-based fighters and only prototypes were flown.

Even more daring was his hybrid jet and rocket SR.53 fighter, which flew in 1957. Alas, that was the year of the infamous White Paper which decreed that the day of manned fighters was over. The SR.53 and a production version, the SR.177 which promised sensational performance, were cancelled.

Gouge, by then chief executive of the company, retired in 1959.

GRAHAM-WHITE, CLAUDE (1878-1958) – early British aviator. Graham-White came to fame when the *Daily Mail* offered £10,000 prize money for the first flight between London and Manchester. Landings could be made en route within an overall time of 24 hours. He had already made one attempt when he tried again on 27th April 1910. This time he had a competitor, **Louis Paulhan**.

Both flew Farman biplanes and took off on the same day making the event a dramatic race, no doubt to the delight of the *Daily Mail*. Public interest was intense. Both pilots landed at nightfall, but Graham-White took off again during the night, an almost unprecedented step at the time. It was to no avail, for engine failure put him out of the race, but he had become a public hero.

Graham-White set up a flying school and established Hendon aerodrome and its regular displays, where the RAF museum now stands. He was an immensely popular figure and commanded huge fees to appear – he was paid $50,000 to fly at an airshow in Boston.

He built aircraft of his own but they failed to win military orders. One, prosaically called the 'Bus', made a brief claim to fame in carrying nine passengers in 1913, a world record at the time. He flew with the RNAS during the war, but in the meantime his works were requisitioned. Disillusioned, he left the country in 1925.

GREEN, GUSTAVUS (1865-1964) – builder of the first British aero-engines. Green built his first engine in 1904, a four-cylinder in-line unit. His engines were used by many British pioneers, including **Moore-Brabazon** for the first flights in Britain by a British national.

Green engines were considered reliable but rather heavy. They were rather outclassed by the end of World War One and he ended his work on them in about 1919.

GRIFFITH, DR. ALAN ('Griff' or 'Soapbubble') (1893-1963) – British gas turbine pioneer. Griffith started on research into materials, earning his 'Soapbubble' name from a 1917 paper on using soap films to investigate torsion.

As early as 1926 Griffith was working on designs for axial-flow turboprops – unlike **Frank Whittle**, he did not see the possibility of a pure jet.

He is regarded as an enigma, sometimes brilliant but often quite impracticable in terms of manufacturing reality. His early work was at Farnborough, but after his initial visionary ideas he let them languish during much of the 1930s. Even worse, he was scornful of Whittle's proposals, particularly his choice of the simpler centrifugal compressor.

In 1939 he moved to Rolls-Royce and designed some theoretically highly efficient but impractically complex engines. **Ernest Hives** employed him as a sort of long-term 'think tank'. He started the early studies which led to the axial-flow Avon, until the task was handed to more practically-minded engineers.

In the 1950s he conceived the idea of 'lift jets' for vertical take-off, both for fighters and for great airliners he sketched which would need no runways. Some prototypes were tested, but the cost and weight of carrying around numerous small engines only used for a few minutes per flight were too much of a handicap. One Griffith 'lift engine', the RB162, did go into service, as a boost engine on a version of the Trident airliner. Ironically it was mounted in the one way he never intended – horizontally!

No Problem

Early in his career Griffith specialised in investigation of metal fatigue until told to drop it because, he was told "there is no fatigue problem in aircraft". Oh dear!

GRUMMAN, LEROY (1895-1982) – founder of the Grumman company. LeRoy Grumman formed his company in 1930, after working with Loening, a make which has long since passed into history. His first design was the 1931 FF1, commonly dubbed 'Fifi' for obvious reasons. It was a naval fighter, and unusually for a biplane had a retractable undercarriage. His factory was widely referred to as the Iron Works.

It was the beginning of a long tradition for Grumman of supplying naval aircraft. His wartime Wildcats, Hellcats and Avengers were among the best carrier-borne aircraft in service. Many were used by the Royal Navy as well as the US Navy. A Bearcat has for some years held the record for the fastest-ever piston-engine aircraft, at 528 mph (850 kph). Both airframe and engine were heavily modified.

After the war, Grumman continued with his line of marine felines. His first jets were the Panther and its swept-wing version, the Cougar. Later came the F-14 Tomcat, first flown in 1970 and still a mainstay of US naval airpower in the 1990s.

Grumman became worried about depending too much on military business and started a range of civil aircraft. First was a most unlikely-looking success story, an agricultural biplane called the Ag-Cat. It sold well, and was followed by a line of fine business aircraft. This division is now a separate company, while the original parent Grumman firm is part of Northrop.

A unique feature fitted by Grumman to his Bearcat fighter was the idea of detachable wing-tips. They were designed to shear off if the wings were subject to high stresses, so relieving the load on the rest of the structure. They did work exactly as planned on at least one occasion, but what would happen if only one sheared off?

GUYNEMER, GEORGES (1894-1917) – leading French 'Ace' of World War One. Between July 1915 and September 1917 Guynemer shot down 54 enemy aircraft. He became a member of an élite group known as *Les Cigognes*, the Storks. Its members became French cult heroes, and the government wanted to take Guynemer off operations and use him for propaganda.

Despite the chance to be rested from operations, Guynemer insisted on staying with his unit. On 11th September 1917 he disappeared on a mission. His fate was never explained. Photographs taken of him towards the end showed drawn and gaunt features, in contrast to the confident look of a few months before. Nowadays it would be recognised that he had stayed on operations too long. Even the bravest of men, and certainly Guynemer was one, are liable to 'crack up' if kept under the strain of constant danger for too long, and it is clear he should have been rested earlier.

HAFNER, RAOUL (1905-80) – rotorcraft pioneer. Austrian-born, Hafner flew an autogyro with controllable-pitch rotor blades in 1927. In 1933 he came to Britain and developed a range of prototypes in the pre-war years which performed well for their time. He was one of the first to experiment with cyclic pitch control, the vital step towards a workable helicopter.

During the war years he devised some bizarre rotary-wing projects. One was the Rotachute, an alternative to the parachute, and another was the Rotabuggy or Rotajeep. This last was a jeep fitted with rotors and a tail. It was to be towed by a tug aircraft, released near an objective, land, discard the rotors and then be driven away. Flight trials ended with some thoroughly frightened pilots – or were they drivers? The idea was superseded by gliders, which in another twist were themselves displaced by helicopters.

In 1944 he was invited to lead a new helicopter division set up by the Bristol Aeroplane Company. He designed the first successful British helicopter, the Sycamore, which first flew in 1947 and gave good service with the British and German armed services.

Unfortunately the good start was not maintained, and Bristol never made another truly successful helicopter. Years of work on twin-rotor machines yielded only the tiny reward of 26 Belvederes for the RAF. The helicopter division merged into Westland.

In November 1980 Hafner and three companions disappeared whilst sailing near Land's End.

No Scoop

Hafner's Sycamore helicopter was being ground run near a marquee where the press were being entertained after seeing the Brabazon airliner. Suddenly the rotors disintegrated with a loud report and some parts landed on the marquee.

A sensational story was feared, but the worries proved unfounded – none of the press noticed!

HAGG, ARTHUR (1888-1985) – British aircraft designer. Hagg joined the de Havilland company at its inception and was jointly or wholly responsible for the company's designs until he moved to Airspeed in 1937.

He is credited with the Dragon and Dragon Rapide biplane small airliners, the similar but larger DH86, and contributed much of the work on the Moth light aircraft.

In a break with de Havilland's traditional biplanes, he designed the DH91 Albatross, an efficient four-engine monoplane airliner of exceptionally pleasing lines. On entry into service in 1938 its flight time from Croydon to Paris was just 53 minutes, not much longer than modern jets. In fact the total journey time was quicker then as check-in was much shorter! However, Hagg's wooden structure was less advanced than his aerodynamics, and the fleet had to be withdrawn during the war.

Hagg invented differential ailerons, since widely used. The 'down' aileron on the outer wing in a turn tends to exert more drag than the other, so Hagg arranged that the 'up' aileron would move further than its partner. It was an important advance in aviation progress.

His best-known product with Airspeed was his Ambassador airliner. It was a fine aircraft, but Airspeed's loss of independence and the coming of turbine power limited sales to 20 for BEA. A twin-engine high-wing design with gently curving fuselage sweeping up to a triple-fin tail, many consider the Ambassador to rival his Albatross among the most beautiful aeroplanes ever built. Hagg seemed to have a particular artistry in creating elegance.

HALFORD, FRANK (1894-1955) – aero-engine designer. An RFC officer, Major Halford was released to help William Beardmore of Dumfries improve an Austro-Daimler engine they were building under licence. The developed version became the BHP (**B**eardmore-**H**alford-**P**ullinger) of 230 hp. It was not really successful until the cylinder heads were redesigned, but once this was sorted out over 4,000 were made, the majority as the Siddeley Puma. It was one of the most important of wartime engines.

In 1923 he became an independent designer, covering cars and motorcycles as well as aero applications. He successfully raced his own cars and motorcycles too. His first aero-engine for light aircraft, the Cirrus, was basically half a wartime design and was intended as an interim product until he could develop his own from scratch. In 1926 the first of his de Havilland Gipsy engines were ready and they proved a world-wide success. Most were 'inverted', with the cylinders at the bottom to give the pilot a better view.

As well as work for de Havilland, he developed engines for Napier, including the powerful but troublesome wartime Sabre.

He started on jet engines with de Havilland in 1941. The first Goblin was run only eight months after the first drawings were issued, and two were fitted for the first flight of the Gloster Meteor on 5th March 1943. De Havilland's own first jet fighter, the Vampire, flew in September 1943 with a single Goblin. Both Vampire and Goblin were widely used throughout the world.

His larger Ghost engine powered the early Comet jet airliners, entering service in 1952. Still generally referred to as Major Halford, he became Chairman of the de Havilland Engine Company when it was created as a separate entity in 1944.

HAMEL, GUSTAV (1889-1914) – pilot of first air-mail service in Britain. Hamel learned to fly in 1910 and within a month had won a cross-country race. He quickly established a reputation as one of the leading aerobatic and racing pilots of his day.

His mail service carried post from Hendon to Windsor. Intended to commemorate the coronation of King George V, it started on 9th September 1911. Hamel flew the inaugural service in a Blériot. It ended on the 26th.

On 17th April 1913, he flew with a passenger from Dover to Cologne non-stop, also in a Blériot. The flight, the first between Britain and Germany, met bad weather and at one point hailstones drew blood on his face.

By 1914 he had made at least 14 Channel crossings and was even planning an Atlantic flight. On 23rd May 1914 he disappeared at sea whilst flying from Boulogne to Hendon in a Morane-Saulnier monoplane.

Rocking the Train

One of Hamel's air-mail pilots found himself flying beside a train. The sight of an aeroplane was still a novelty and he saw crowds of faces at the windows. To give those on the other side a chance, he flew over the line and realised all the passengers had crossed to that side. Mischievously he then flew from one side to the other until the train rocked so much that the driver stopped to investigate.

HANDLEY PAGE, SIR FREDERICK (1885-1962) – one of the great names in British aircraft building. 'HP' started his long career in aircraft construction with his HP1 *Blue Bird* of 1909. It was a monoplane with crescent-shaped wings which gave it an appearance well ahead of its time. He would return much later to crescent wings, albeit with the curve in the opposite direction, in the Victor V-bomber of the 1950s.

For much of his lifetime, Handley Page specialised in large aircraft. At one time any big aeroplane tended to be called a 'Handley Page', often to the annoyance of other firms. His 1915 O/100 twin-engine bomber and improved O/400 were among the first strategic bombers. The larger O/1500 four-engine bomber was designed to reach Berlin, but the Armistice was signed just before its first mission.

In 1919 he started an airline, Handley Page Air Transport, using adapted bombers. Services were short-lived – enthusiasm could not overcome the poor economics of the aircraft then available. They made one advance in air travel in being the first to serve meals in flight.

In post-war years times were tough for aircraft builders, but 'HP' astutely bought up quantities of war-surplus machines and sold them under the name Airdisco (**A**ircraft **D**isposal **C**ompany). It helped to keep him solvent at a difficult time.

Always keen to make flying safer, he achieved a significant reduction in take-off and landing speed with his Handley Page leading-edge slots. Licences were sold all over the world and thousands of lives must have been saved thanks to this invention.

On holiday in Belgium he showed his concern for safety in another way. A big man, he managed to rescue two people from drowning, a feat for which he was decorated by the King of the Belgians.

He continued developing bombers and airliners. The best known were the four-engine biplane airliners, the HP42 *Hannibal Class*, first flown in 1932. Their antiquated appearance may have been something of a joke, but their safety record was not – they never killed a passenger throughout their airline service.

As war threatened again, he produced the Hampden medium bomber and, more importantly, the four-engine Halifax, one of the principal British heavy bombers. Over 6,000 were made.

After the war came the impressive Victor jet bomber, considered by many to have been the best of the V-bombers. By then political pressures were bearing on 'HP', for the government wanted all the aircraft firms to merge into larger units. The independent-minded Sir Frederick held out for what he thought to be a fair deal, only to see orders for his Victor cut back and a potential RAF purchase of his Herald transport diverted to Avro. The firm survived its founder by just seven years.

One of the 'greats' of British aviation, 'HP' was a towering personality known for his vast fund of stories and prodigious memory – he had a remarkable knowledge of the bible and could select a quotation for almost any occasion.

'HP' was an indifferent pilot, before he devoted all his time to manufacturing. After one painfully bouncy landing he asked a friend "Did you see my landing?" "Yes", came back the answer, "I saw them all."

HARGRAVE, LAWRENCE (1850-1915) – Australian pioneer of kites, aerofoils and rotary engines. Hargrave was born in Britain, but all his work was performed in Australia.

He experimented with aerofoils from 1883 onwards and made some significant observations about profiles and lift characteristics. In parallel, he examined the problem of power for flight and has been credited with the invention of the rotary engine, which came into its own just before the First World War.

In the 1890s he invented the box-kite, and lifted himself off the ground with a string of kites. The box-kite formed the basis for early European aircraft, as well as still making a popular toy.

HARRIS, SIR ARTHUR (1892-1984) – wartime leader of Bomber Command. A former RFC pilot, Harris took over leadership of Bomber Command in 1942. He believed the war could be won by bombing, following **Douet's** theory, and concentrated his resources on area bombing of cities.

The heavy aircrew and German civilian casualties of his approach have made it controversial ever since. He was by no means single-minded in his policy and did authorise some precision attacks on strategic targets, like that on the rocket research establishment at Peenemünde.

As the plans for the Normandy invasion advanced, progressively more of his effort had to be diverted to tactical targets.

In some respects ruthless, he was far more sensitive to casualties than sometimes portrayed. Commonly referred to in the press as 'Bomber Harris', this name was not used by crews, who generally held him in high respect. To them he was 'Bert' or 'Butch' Harris. He retired in 1946.

The Bombing Campaign – Right or Wrong?

It is not easy to answer the question as to whether the heavy losses among RAF crews and victims on the ground were justified. There was enormous pressure from the British civilian population to 'hit back', and bombing seemed the only way available.

From a military viewpoint, heavy damage was caused to munitions production, and large resources had to be put into nightfighter defences. As the Nazis were killing several hundred thousand people a month in their extermination camps, if the bombing advanced the end of the war by just a couple of months it must have saved more lives than it ended.

Most surviving aircrew defend Harris' policy fiercely. This is despite the 47,000 bomber aircrew killed in the war, 70% of all RAF losses.

Stuart Harrison of the US Navy was approaching to land, fortuitously a little too high, when the engine of his F-8 Crusader flamed out. Instead of ejecting he landed without power on the deck, making probably the only 'dead-stick' landing in carrier history. Think about it!

HARTMANN, ERICH (1922-93) – most successful fighter pilot of all time. The score of 352 victories credited to the German pilot Hartmann has never been equalled in the history of aerial warfare, and probably never will be.

By no means a reckless pilot, Hartmann built up his score slowly to start with, whilst gaining experience. Later on, he was to shoot down as many as 78 aircraft over a four-week period. Most of those were Russian opponents, some of whom, however determined and courageous, had been given insufficient training, and many of his victims were vulnerable transports.

He had a reputation for remarkable accuracy in 'deflection shooting', anticipating the movement of the other aircraft. It is a mark of his skill, as well as a measure of luck, that he was only forced down twice himself.

He beat the odds and survived the war and ten years imprisonment in the Soviet Union before his release.

Flying for the Fatherland

Some readers might feel that relating the high scores of pilots like Hartmann is glorifying war or justifying the Nazi regime. This is far from the case. They must be included for historical reasons, but in any case few of the German pilots had any idea of the atrocities committed by the Nazis.

Like young men everywhere, they believed they were serving their country with honour.

HAWKER, HARRY (1889-1921) – test pilot who gave his name to the Hawker company, An Australian, Hawker broke several height and endurance records before the First World War. He joined Sopwith as Chief Test Pilot, but he was also involved in design.

He proved all the company's wartime products such as the Pup, '1½ Strutter', Camel and Triplane. He looped the last-named within three minutes of starting its maiden flight. Such flying would be branded irresponsible today, but not in the climate of the time.

In 1919 he made a bid for the first Atlantic flight in the Sopwith Atlantic, a biplane specially designed for the flight, as its name suggests. He was accompanied by Kenneth McKenzie-Grieve as navigator. During the flight the engine overheated badly, possibly due to a misleadingly labelled radiator shutter control. As the engine seemed likely to seize, they decided to ditch if they saw a ship, and they were remarkably fortunate to do so.

The small Danish vessel *Mary* was slow and lacked radio, so by the time they reached land a week later the two airmen had been given up for lost.

In 1920 the Sopwith company succumbed to the loss of wartime output and ceased trading. Some of its former managers reformed it under the Hawker name, but with Tom Sopwith remaining as chairman.

It later built such famous aircraft as the Fury, Hurricane, Typhoon, Hunter and Harrier, but Hawker, despite giving his name to the company, was not to see this.

Despite his epic flights, Hawker was not physically robust. In 1921 he crashed at Hendon. It is believed he died of a heart attack in the air, but it was never proven and there were alternative theories involving technical faults with the aircraft.

HAWKER, LANOE (1890-1916) – one of the first air Victoria Cross holders. Hawker flew 'scouts', the predecessors of fighters, in the early part of the war. He put teeth into his missions using a revolver and small bombs or grenades, but he wanted more effective armament.

In desperation he made a rudimentary fighter by attaching a deer-stalking rifle, and later a machine-gun, on the structure and angled to avoid the propeller. Courageous and ruthless, he is credited with 73 victories. He was awarded his VC in August 1915.

He used his experience to improve flying and combat training. He became a unit commander and as such was forbidden to fly over enemy lines. He was not a man to observe such a constraint, and in November 1916 he fell prey to **von Richthofen** after a long engagement which left him at a disadvantage due to lack of fuel.

HAWKINS, WILLIS – designer of the Lockheed C-130 Hercules. The needs of the Korean War showed an urgent requirement for larger and faster military transports than any then in service. Willis designed the C-130 under the general guidance of **'Kelly' Johnson**.

He chose a high wing layout with a cavernous fuselage loaded through a large rear ramp. The power was provided by four Allison T56 turboprops.

Most unusually, the C-130 beat its specification in every respect. It has been in production ever since its first flight on 23rd August 1954, probably making it the longest production run of any Western aircraft.

The Labours of the Hercules

The C-130 Hercules has to be one of the most versatile aircraft ever built. 64 countries (so far!) have bought the type and new versions are planned which should take production beyond 50 years.

Wherever relief is needed from famine, earthquake, or any other natural or man-made disaster, almost certainly succour will come in the great cargo holds of the Hercules. Its ruggedness and ability to land in tight spots are legendary.

There was even a formation display team flying the Hercules at one time, although their claim to be an 'aerobatic team' was stretching matters, and believe it or not, trials were made from an aircraft carrier.

HAZELDEN, HEDLEY (1915-) – Handley Page test pilot. After wartime service with Bomber Command, Hazelden joined Handley Page as a test pilot. He handled much of the Victor test flying. Just before one test, a last-minute engagement caused another pilot to take the flight, during which the tailplane failed and all aboard were killed.

His best-known incident occurred when flying the prototype Herald airliner to the 1958 Farnborough Show. It had shortly before been fitted with Rolls-Royce Dart engines, replacing original piston units, and one of these normally most reliable engines developed a major fire.

It was only by the greatest of skill that he was able to land in a far-from-ideal field without injury to any of the occupants.

HEINEMANN, EDWARD ('Ed') (1908-91) – Designer of Douglas military aircraft. Many of Heinemann's designs became classics, remaining in service long after their original life expectancy. He started with Douglas in the late 1920s, left to join Northrop, then following a merger between the companies found himself back with Douglas.

Among his best wartime types were the A-20 Boston twin-engine bomber and the Dauntless naval attack aircraft. After the war he produced the first aeroplane to exceed Mach 2, the Douglas Skyrocket. The milestone was passed on 20th November 1953.

His 1954 A-4 Skyhawk, often called 'Heinemann's Hot-Rod' or 'Tinkertoy', proved one of the finest combat jets ever made. Most aircraft turn out heavier than intended, but Heinemann managed to build it at only half the weight specified by the US Navy, an almost unprecedented achievement in aviation!

Almost 3,000 were made over 25 years and it proved its effectiveness in action with American, Israeli and Argentinian pilots in various wars.

HEINKEL, ERNST (1888-1958) – German aircraft constructor and builder of first ever jet aircraft. During World War One Heinkel worked for other companies before forming his own in 1922. He built some biplane fighters, but his 1932 monoplane He 70 was a trendsetter, showing some of the cleanest lines then seen on any aeroplane. Rolls-Royce bought one as a flying test-bed thanks to its high performance.

Less welcome to the British people was its twin-engine derivative, the He 111 bomber. The sound of its throbbing, unsynchronised engines came to be hated throughout Europe.

Heinkel backed jet pioneer **Hans von Ohain** in developing his early engines. On 27th August 1939 the Heinkel 178 made the first flight in the world by a jet aircraft. Fortunately for the Allies, there was as much official dithering over jets in Germany as there was in Britain.

Heinkel, Ernst *(continued)*

Ernst Heinkel had made enemies among top Luftwaffe officers and his jet projects were given low priority until late 1944. Then in September of that year he was cleared to design a jet fighter. Almost unbelievably, his team had their He 162 Volksjäger (People's Fighter, also called Salamander) flying by the end of the year. It was too late to have much effect on the war, and it was too demanding to fly for the inexperienced youth pilots for whom it was earmarked.

It was also Heinkel's unpopularity which limited production to small numbers of what was probably Germany's best nightfighter, his He 219 Uhu (Owl).

HENSON, WILLIAM SAMUEL (1812-88) – visionary of air travel. The British inventor experimented with gliders and steam-powered models, some with **John Stringfellow** as partner.

In 1842 he proposed a steam-powered airliner and drew fanciful drawings of his *Aerial* serenely cruising over distant countries. Rashly, he even announced the founding of an airline. Although he over-reached himself with ambition and exposed himself to ridicule, in some respects his design was truly prophetic. Its monoplane cambered wings, enclosed cabin and tricycle undercarriage were remarkable visions of the future.

Henson and Stringfellow made a steam-powered model in 1847, but it failed to fly. He emigrated to America the next year, supposedly disillusioned by public reaction to his scheme.

HENSHAW, ALEX (1912-) – record-breaker and test pilot. Henshaw was one of the outstanding racing and record-breaking pilots of the 1930s. Racing at the time was hazardous when engines and equipment were less reliable than they are now. When his engine failed over the Irish Sea during a race he was fortunate to have been able to ditch near a ship.

In 1939 he broke the solo record to Cape Town and back in a Percival Mew Gull. The return trip took 4 days 10 hours, of which 78 hours was spent in the air. It was an extraordinary feat of airmanship and endurance and proved the most durable of all flying records. Few pilots today would accept the risks he took in landings at primitive airfields in bad weather and battling against fatigue. Once he was about to enter a dark cloud in poor visibility when he realised at the last moment it was a mountain. He had to turn so hard that he blacked out momentarily. Nowadays the flight would doubtless be thought foolhardy but attitudes towards risk were different in 1939 and he became a public hero.

He joined the team test-flying the Spitfire. His low-level Spitfire aerobatic displays became legendary. His skill was recognised in his appointment as Chief Test Pilot at Castle Bromwich when that plant produced Spitfires and bombers.

His team conducted 37,000 wartime test flights, pilots sometimes testing over 20 aircraft in a day. Inevitably there were incidents, and he was lucky to survive one accident when his Spitfire was wrecked in a forced landing among houses. He was none too pleased when local gunners once added to the hazards by opening fire on him.

The Welcome Gift

Whilst practising aerobatics in a light sports biplane, an engine failure and serious fire made Henshaw grateful for a recent present from his father - a parachute!

HESS, RUDOLPH (1894-1987) – Nazi leader who flew to Scotland in 1941. The mysterious flight of Hess in a Messerschmitt 110 fighter to the Duke of Hamilton's estate has puzzled aviation and political historians for decades. He baled out rather than risk a night landing.

His motives have never been publicly revealed. Whatever they were, the effort did him little good, for he spent the rest of his life in prison, for the last few years as the sole occupant of Spandau.

HILL, GEOFFREY (1895-1956) – designer of tail-less aircraft. Hill was a wartime pilot with the RFC, then a test pilot at Farnborough. He became attracted to the idea of tail-less aircraft and Westland agreed to build research aircraft based on his designs. These were the Pterodactyl family.

The first of the 'Pterrible Pterodactyls', which owed something to **John Dunne's** work, flew in 1928. Pictures taken from below show a remarkable resemblance to modern microlights – Hill was 50 years ahead of his time.

A series of prototypes was built, but while they showed some promise they were unstable in pitch and tricky to fly. None went into production.

In the 1950s another research aircraft was built under his direction, the Short Sherpa, with movable wing-tips to relieve stresses on thin swept wings. He called it the aeroisoclinic wing, but it was not taken further.

HINKLER, HERBERT ('Bert') (1892-1933) – made first solo flight from Britain to Australia. Having served first as a gunner then as a pilot in the First World War, Australian-born Hinkler joined Avro as a test pilot in 1920.

In that same year he attempted to fly to his homeland in a tiny, and to most people quite unsuitable, Avro Baby with a puny 35 hp engine. Courage he never lacked. At Turin the authorities refused him permission to continue.

Hinkler, Herbert *(continued)*

After several years test flying and racing with success, he set out in 1928 on a second bid to reach Australia, this time in an Avro Avian. He reached Darwin 15½ days later, completing the first solo flight between the countries.

Following an unsuccessful dabble in aircraft design, Hinkler returned to the record trail. Among his notable flights was the first solo crossing of the South Atlantic in 1931, flying a de Havilland Puss Moth.

By 1933 his time to Australia had been beaten and he sought to retake it. His Puss Moth disappeared in bad weather among the Apennines. His fate was unknown for three months, until the wreckage was found by chance.

A quiet and modest man, Hinkler was regarded as one of the greatest pilots of his day.

HIVES, ERNEST (Lord Hives) (1886-1965) – led Rolls-Royce through the war and into the turbine age. As a garage mechanic, Hives offered to sort out a fault on a passing motorist's car late one evening in 1908. The motorist was **Charles Rolls** and that chance meeting led to a nearly 50-year association with Rolls-Royce.

Hives tested and raced the company's cars, and soon became head of the Experimental Department, building prototype cars and later aero-engines.

He played a major part in preparing the Schneider Trophy engines in 1929 and 1931. The frenetic pace of development is shown by the rise in power from a 925 hp engine to an output of 2783 hp from essentially the same unit. So stressed were the engines that failures were inevitable, and when a problem was found the evening before the race Hives winkled fitters out of cinemas and bars to work throughout the night. The racing 'R' engines held the world speed records on land, sea, and air, though never quite at the same time.

In 1936 Hives, or 'Hs' in the company's shorthand naming system, became a member of the board and effectively led Rolls-Royce through the demanding wartime years. Not only did he oversee the development of the Merlin and Griffon to cope with pressure of ever-greater powers, but he also enormously expanded production by building new factories at Crewe and Glasgow, and later arranging manufacture at Ford in Britain and Packard in America.

When Britain's jet engine programme was stagnating at Rover, Hives in his decisive manner offered a solution: he would take over Rover's jet factory in exchange for the Rolls-Royce tank engine plant, which built a version of the Merlin called the Meteor. The deal was swiftly done and Rolls-Royce became as much a leader in turbine engines as it had with 'thumpers'. Hives guided development of the early jets such as the Derwent and Nene, and later the technically demanding axial-flow Avon.

He also put the company firmly into the airline business with the Dart turboprop, later a world-wide success, but at first a cause of despair when it was overweight and below power.

He declined a wartime knighthood on the grounds that honours should be confined to those doing the fighting, but to popular acclaim was awarded a barony in 1950.

Universally respected and a worthy successor to Royce, Hives retired in 1957.

Merlin Magic

The Merlin, the most famous of aero-engines, started as a private venture i.e. without government support. Towards the end of his life, Henry Royce approved its launch, doing his country an immeasurable favour. Like most of the company's piston engines, it was named after the bird of prey and not the magician.

At first the Merlin gave plenty of trouble, but eventually it went on to power vital front-line aircraft such as the Hurricane, Spitfire, Mosquito, most Lancasters, the Mustang and many more. Under Hives' skilled guidance its power was more than doubled during the war.

Over 166,000 were made. He was keen to see it put the company into post-war airline business, and although not designed as a civil engine it gave good service in types such as Yorks and the Canadair Four (Argonaut).

Ironically the last military Merlins flew in its old foes – Spanish-built versions of the Messerschmitt Bf109 and Heinkel 111.

Beyond description in print is the sound of the Merlin. It is purposeful, it is nostalgic, it is unique, and it can move older men to tears.

Lord Hives was aboard an airliner which developed an engine problem. Nearby passengers asked him what might be wrong with the engine. The great proponent of in-line and Vee engines glanced at the offending radial and replied dryly "It's the wrong shape."

HOBBS, LEONARD ('Luke') (1897-1977) – designer of Pratt and Whitney aero-engines. Hobbs joined the American engine maker in 1927. He was responsible for the company's range of twin-row radial engines which became among the most widely used of all aircraft power units.

Wing Commander 'Taffy' Holden flew a Lightning jet fighter – by mistake! He was doing taxi tests to diagnose an electrical fault when he accidentally engaged the afterburners on the two Rolls-Royce Avons. Before he found out how to cancel them he was at take-off speed.

Never having flown anything faster than a Chipmunk, he found himself airborne in a Mach 2 fighter without canopy, radio to seek advice, or an ejection seat. After two missed approaches he managed to end his terrifying ordeal with some semblance of a landing on the third attempt.

HOOKER, SIR STANLEY (1907-84) – aero-engine designer. A mathematician, Hooker came to prominence in greatly improving the performance of Rolls-Royce Merlin superchargers. His work made a major contribution to increasing the power wrung from the engine and so improving the speed of British combat aircraft.

When Rolls-Royce took over jet engine work from Rover, **Ernest Hives** placed Hooker in charge of development at Rover's former works at Barnoldswick. There he saw the Derwent into production and developed the Nene. This engine may have been the most geographically widely used aero-engine of all time, having been built in Britain, France, Australia, the United States, Canada, and in all but name in Russia and China too.

After a row with Hives he moved to Bristol and sorted out the then troublesome Proteus for the Britannia, and developed the Olympus and the highly successful smaller Orpheus.

In the 1960s he worked with **Sir Sydney Camm** to design the revolutionary Pegasus vectored-thrust engine for the Harrier vertical/short take-off combat aircraft. The concept owed a little to ideas of a French designer, Michel Wibault.

Hooker retired in 1967, or so he thought. In 1970 he was summoned to put the big RB211 programme in order. It was in deep trouble, largely as a result of the sudden death of Chief Designer Adrian Lombard. He did resolve the problems, but not before the engine had dragged the company and nearly Lockheed as well into insolvency.

Eventually his work proved so effective that the engine became one of the most reliable, powerful and efficient in airline service and a huge export earner for Britain. Incidentally, the 'RB' in the type numbers still comes from his old haunt in the former Rover works – 'Rolls-Royce Barnoldswick'.

Lady Houston was not directly an aviation personality but one act of her generosity and patriotism did much to ensure that Britain had high-performance combat aircraft ready by 1939. When the government refused to fund a bid for the Schneider Trophy in 1931, Lady Houston stepped in with a £100,000 donation.

The pressures of the race gave a huge impetus to airframe, engine and fuel development. Reginald Mitchell, designer of the Spitfire, believed it advanced such development by three years.

HUGHES, HOWARD (1905-76) – record-breaker, aircraft builder and eccentric. Heir to an industrial empire, Hughes entered aviation in the 1930s. He produced the classic film *Hell's Angels* in 1930 about First World War flying. When sound pictures came in he made it again.

He set the world speed record in 1935 at 353 mph (560 kph) in his efficient H-1 racer of his own design. He went on to break the American coast-to-coast record in 1936, and in a Lockheed 14 broke the round-the-world record in 1938.

During the war he undertook construction of a giant flying-boat capable of carrying 700 troops. Its 320 ft (98 m) span still makes it the largest aeroplane ever built. Originally called the HK-1 Hercules, later renumbered H-4, and popularly (but not by Hughes) the *Spruce Goose,* it made just one flight.

Before that, in 1946 Hughes was seriously injured in a crash while testing a new fighter, his XF-11. He almost lost his life and it is probable his personality changed from that time. Even so, he was well enough to fly his colossus in November 1947 when he lifted it off the water for about a mile. It never flew again and has been preserved ever since as a tourist attraction.

Hughes gained control of the airline TWA, but became progressively more eccentric and lived his last years as a recluse.

ICARUS – mythical son of **Daedalus**, supposedly escaped from the Labyrinth by air. Alas for Icarus, he failed to do his thermal calculations and the sun melted the wax attaching his wings.

ILYUSHIN, SERGEI (1894-1977) – Russian aircraft designer. Ilyushin learned to fly in 1917. His most successful design was his Ilyushin 2 Shturmovik (Storm), a single-engine ground attack aircraft. Where Ilyushin scored with this aeroplane was in designing it from the outset with plentiful armour plating to protect it from ground fire. At first production was slow, so much so that it incurred Stalin's anger.

Most people who did that disappeared, but the output soon improved, so much so that 36,163 were eventually made, more than for any other World War Two type. Ilyushin evidently regained Stalin's favour, for the dictator described the aircraft "As important to the Red Army as air or bread."

After the war Ilyushin built many of the most widely-used Soviet airliners and bombers.

IMMELMANN, MAX (1890-1916) – fighter pilot and originator of the Immelmann Turn. Far from a natural pilot, Immelmann was at first notorious for his clumsy flying, but he became one of the first true fighter pilots. He devised the Immelmann turn, in which he dived below an enemy then climbed and fired from below. He then pulled up almost vertically, applied hard rudder and dived back upon the victim from the new direction. It must be said that there are different versions of what he really did!

Immelmann became one of the leading German combat pilots. He was killed on 18th June 1916 when his aircraft broke up, possibly due to failure of the interrupter gear.

IRVIN, LESLIE (1895-1966) – pioneer of the rip-cord parachute. An American, Irvin made his first jump from a DH4 at McCook Field, Dayton, on 19th April 1919. Shortly after, he was given his first order for 300 parachutes. The first emergency use of an Irvin canopy was by Harold Harris in 1922.

In 1926 he came to Britain and founded the company which has been a leading supplier of parachutes ever since. Although an American, he was fiercely patriotic on Britain's behalf during the war years.

His products have saved some 32,000 lives. Few men have made a greater contribution to humanity in the air.

An Exclusive Club

The Caterpillar Club has just one entry qualification – a member must owe his or her life to an emergency jump using an Irvin parachute.

When Leslie Irvin formed his company in Britain the registrar made a slip and it was set up, mis-spelt, as the Irving Air Chute Company. The mistake was left uncorrected and remained a historical curiosity.

IRWIN, CARMICHAEL ('Bird') (1894-1930) – commander of the airship R-101. After serving in non-rigid airships from 1915, Irwin became one of the most experienced airship commanders. In the sporting field, he ran in the 1920 Olympics.
He handled much of the R-101 testing and commanded her on the final flight. The disaster happened only minutes after he had handed over the watch, suggesting he was confident in the airship despite stormy weather. The final chain of events , possibly starting with a section of envelope coming adrift in the rain, must have developed very rapidly.

JATHO, CARL (1873-1933) – German aeroplane pioneer, made short hops in 1903. Jatho flew a small biplane 59 feet (18m) on 18th August 1903. With a monoplane he flew 196 feet (60 m) in the autumn of the same year.
Although these trials preceded those of the Wright Brothers, they are not considered to have been sustained or controlled, the essential qualifications for true flight.

JOHNSON, AMY (1903-41) – British long-distance woman solo pilot. Just 80 hours after her first solo, Amy Johnson took off for a heroic solo flight to Australia in a second-hand Gipsy Moth *Jason*. Nineteen days later, on 24th May 1930, she confounded the pundits by arriving safely at Port Darwin.
She followed this epic with an attempted flight to Tokyo, crossing Russia in winter. Despite her determination, she came to realise that an open cockpit and a Siberian winter were an unhappy combination. She abandoned that flight, but a few months later she was back again, this time in an closed-cockpit Puss Moth, and reached Tokyo in a record 9 days.
In 1932 she married another record-breaker, **Jim Mollison**. Perhaps undiplomatically she broke his solo record to Cape Town later that year, and for good measure did it again in 1936. They made several notable flights together, but an attempt on the long-distance record ended in a forced landing and serious damage. There was reputed to be as much turbulence within the cockpit as outside it and divorce followed in 1937.

Johnson, Amy *(continued)*

During the war she joined the Air Transport Auxiliary, ferrying aircraft to and from operational bases. In 1941 she disappeared whilst flying an Airspeed Oxford. She was seen to parachute into the Thames Estuary. A naval officer, Lt Cdr Fletcher, almost rescued her but both were drowned. There was speculation that she might have been accidentally hit by British guns, but most opinion now is that she became lost and ran out of fuel.

JOHNSON, CLARENCE ('Kelly') (1910-90) – Lockheed designer. Kelly Johnson first made his name by designing the twin-fin tail unit on the pre-war Electra airliner, turning it into a far better aircraft and establishing the company's reputation. He may, indeed, have saved the company.

His principal wartime design was the P-38 Lightning fighter, with an unusual twin-engine, twin tail-boom layout. It achieved widespread success in the Pacific, but less so in Europe where its mediocre manoeuvrability placed it at a disadvantage. In the meantime he was also working on the Constellation airliner, one of the most elegant ever made and one of the principal long-haul commercial aircraft for the next decade.

Towards the end of the war he designed the Lockheed P-80 Shooting Star, the first truly successful American jet fighter. It just missed wartime action.

In the 1950s he headed a top secret section of the Lockheed company widely known as the 'Skunk Works'. From it emerged the U-2, looking like an outsize sailplane and ostensibly for high-altitude research. In 1960 the 'research' caused an international outcry when a U-2, flown by Francis Gary Powers, was shot down over Russia and its true purpose became only too obvious.

Johnson's most spectacular creation was his SR-71 'Blackbird', the Mach 3 successor to the U-2 and still one of the most impressive aeroplanes ever made. He retired in 1976.

JOHNSON, JAMES ('Johnnie') (1915-) – top-scoring British fighter pilot of World War Two. Johnson accounted for 38 enemy aircraft, the highest confirmed figure for any British pilot in that war. Interestingly, he did not go into action until 1941, so he achieved this distinction without flying in the Battle of Britain.

He was given command of a squadron in 1942, and the Canadian wing in 1943. Johnson remained in the RAF until 1965, retiring as an Air Vice-Marshal.

JUNKERS, PROFESSOR HUGO (1859-1935) – German aircraft and engine builder; pioneer of metal construction. The Junkers J1 of 1915 set the pattern in cantilever (i.e. without bracing struts or wires) metal construction. He followed it in 1919 with his J13 single-engine four-seat monoplane, which is considered a landmark in airliner design.

Its corrugated metal construction, enclosed cabin, and passenger seat-belts (another 'first') put it years ahead of its time. 322 were built, some fitted with skis or floats, and with their durability proved unequalled in remote areas of the world.

Junkers continued with his strong, but rather 'draggy' corrugated metal structures. The 1930 J38 four-engine airliner was a giant of its day. Some passengers sat in the wing looking forwards, giving an unrivalled view but perhaps unsettling if a mishap seemed imminent. More successful in sales terms was his J52/3M three-engine airliner. It was a mainstay of pre-war Lufthansa and thousands provided basic transport for the wartime German forces. Its success made Junkers the largest aircraft company in the world in the late 1930s.

Hugo Junkers opposed the Nazis and like others who refused to practice easy acquiescence, he paid the price. **Erhard Milch,** whom Junkers had earlier helped in his career, sought to take over the Junkers company. The elderly Hugo Junkers was arrested and died soon afterwards. The way was left open in his absence for the firm to build the formidable Ju88 light bomber and the notorious Ju87 Stuka dive-bomber with its wailing siren to demoralise those below.

KINDELBERGER, JAMES ('Dutch') (1895-1962) – President of the North American Aviation company. After working for Martin then Douglas, Kindelberger became President of the predecessor of North American in 1934.

With designer Raymond Rice he produced some classic American aircraft. They included the B-25 Mitchell, thought by many to have been the best American wartime medium bomber, and the P-51 Mustang. This was produced in response to a request from the British Purchasing Commission. Rice produced an outline design with an efficient laminar wing, and orders were placed in May 1940. By an astounding effort, the prototype flew on 26th October 1940. Early versions used an Allison V-1710 engine but performance at height was uninspiring. Rolls-Royce test pilot Ronald Harker flew one and suggested it would be an outstanding fighter with a Merlin. It was soon done and the Merlin-Mustang is considered by many to have been the best fighter of the war. Where it excelled was in range, allowing it to escort bombers all over Germany.

Other excellent aircraft produced under Kindelberger's management included the F-86 Sabre, America's first swept-wing fighter and the first production aircraft capable of exceeding Mach 1 in a dive. It played a vital role in Korea. Its successor, the F-100 Super Sabre, after early problems, also became one of the leading front-line types of its day.

KINGSFORD-SMITH, SIR CHARLES (1897-1935) – Australian trail-blazer, first to cross the Pacific. With fellow Australian Charles Ulm, Kingsford-Smith started his impressive sequence of long-distance flights with a circumnavigation of Australia in 1927. The journey, in a Bristol Tourer, took ten days.

In 1927 they bought a second-hand Fokker VII tri-motor which they named *Southern Cross*. They flew it, with two other crew, on the first Pacific crossing by air, leaving Oakland on 31st May 1928. After intermediate landings at Hawaii and Fiji, they reached Brisbane in 9 days.

As was usual with such flights at the time, the heavily-loaded take-offs were made with minimal margins and were a major source of worry. They had their share of atrocious weather too. Trailblazing was not for the faint-hearted.

He made further record flights, later switching to single-engine types. He flew an Avro Avian *Southern Cross Junior,* a Percival Mew Gull *Miss Southern Cross*, and a Lockheed Altair *Lady Southern Cross.*

It was a hazardous life and the odds caught up with him in November 1935. Flying the Altair with Tommy Pethybridge, he disappeared over the Bay of Bengal whilst attempting another Britain to Australia record. His name is perpetuated in Sydney's international airport.

LADDON, ISAAC ('Mac') – American aircraft designer. Among Laddon's creations were three outstanding types, the PBY Catalina, the B-24 Liberator, and the ConvairLiner piston airliners.

The PBY, known in RAF service as the Catalina, was the most successful marine aircraft of all time – around 3,300 were made, including some in Canada and Russia.

The Liberator was America's best wartime heavy bomber. Over 18,000 were made in an extraordinary feat of production – around 17 ***a day*** at its peak.

The post-war Convair 240 and its improved versions, the 340 and 440, were sturdy twin-engine airliners. Many were later converted to turboprop power. It is a tribute to Laddon's work that one major airline president described the ConvairLiners as the finest airframes ever built.

Some substantiation of that claim may be found in the record of 150,000 flights logged by one of the Convairs, believed to be the highest total for any aircraft.

LAKER, SIR FREDDIE (1922-) – flamboyant airline head. A flight engineer with the Air Transport Auxiliary, and a qualified pilot, during the war, Laker set up an engineering and trading company for military surplus aircraft and parts in 1947, Aviation Traders. Some of the freighters he owned were put to profitable use on the Berlin Airlift and Freddie Laker became financially secure.

In 1951 he added the airline Air Charter to his empire, eventually forming British United Airways after a series of mergers.

In 1962 he started the world's first public hovercraft service, across the Dee estuary.

As head of British United he did much to launch the BAC One-Eleven airliner, starting off the order book with a purchase of ten.

His engineering company designed and built the Carvair, a conversion of the Douglas DC-4 with a front opening door and cockpit mounted above, Boeing 747 style. It was intended as a car ferry or freighter, and small numbers entered service.

He also built a prototype of a twin-Dart powered business aeroplane, the Accountant, but did not put it into production.

He left British United after a boardroom row and started Laker Airways in 1966. In 1977 he realised an ambition, which had long been thwarted by bureaucratic obstruction, by launching his trans-Atlantic *Skytrain* service. This was a 'walk-on' service without frills – the passenger walked up to the terminal without a ticket and awaited the next flight. By such means as cutting out meals the fares were kept below those of the national airlines.

Due to a combination of over-expansion and (as Laker claimed) cartel-type action by the major carriers, Laker Airways went into liquidation in 1981.

The irrepressible Sir Freddie was not the sort of man to retire quietly, and in 1995 he started afresh with yet another new airline.

A Grave Incident

Freddie Laker was fortunate to survive an accident when he was in the cockpit of a York, although not the pilot in command.

All four engines cut on the approach to Hamburg and it was his quick thinking which avoided what had looked an almost certain arrival in a cemetery.

LANCHESTER, FREDERICK (1868-1946) – pioneer of aerodynamic theory. Lanchester was studying the potential of the petrol engine for both cars and flight in the early 1890s. By 1892 he was extending his study of flight with a model glider. In 1894 he progressed to a rubber-powered model aircraft. Within another three years he had completed a monumental 900-page work, *Aerial Flight,* although it was to be a further 10 years before he published it. His grasp of aerodynamics, power requirements, structures, piloting technique and meteorology was astounding for his time.

During the next decade he concentrated upon cars and power-boats. His cars were sold in respectable numbers and led the way in numerous features which later became commonplace. They are much sought after by collectors today.

In 1907 he at last published *Aerial Flight* in two parts, Aerodynamics and Aerodonetics. It is interesting, if futile, to speculate on whether he could have beaten the **Wrights** into the air had he concentrated on aviation.

In 1911 he built a full-size aircraft with a partner, Norman Thompson. It was one of the first with twin engines. It was designed to meet a military requirement. On its only attempted flight the undercarriage collapsed. Lanchester blamed a late change in the customer specification in adding a second seat, possibly combined with an unsuitable beach used for the trial. It would not be the last argument in the history of aviation between builder and customer over specification changes.

During the First World War, Lanchester acted as a government adviser on aero-engine development and air fighting tactics. Afterwards he returned to vehicle research and studied fields such as electronics and even relativity. He was one of the most remarkable engineers of his time.

LANGLEY, DR SAMUEL (1834-1906) – attempted powered flight in 1903. Langley built a series of steam-powered models from 1986 onwards. At the time he was head of the Smithsonian Institute. Flights of up to 4,200 feet (1,280 m) were achieved, the first sustained powered flights in history, although by models only.

In 1902 he switched to petrol engines. Good progress was continued on models, and in 1903 he moved on to building his full-size *Aerodrome,* with a remarkably efficient 55 hp radial engine designed by his colleague Charles Manly.

On 7th October and again on 8th December the machine was launched from the top of a houseboat on the Potomac River, with Manly aboard. Each time it caught on the launching mechanism and plunged into the water. On the second occasion the *Aerodrome* was considered beyond repair. Like other gallant pioneers who failed, Langley was subjected to merciless ridicule. One newspaper forecast that it would take between one and ten million years before men flew with wings. Actually it would take just nine days.

Many felt that with better luck Langley's machine could have flown, but a later attempt by **Glenn Curtiss** to prove the point by repairing and flying the *Aerodrome* ended in recrimination when it was found he had altered it substantially.

Even if Manly had managed to stay airborne, Langley had provided no real means of control, relying wholly on stability built into the machine – how would he have managed a landing without crashing?

Langley was a talented if self-taught scholar but he slipped up on his Greek when selecting the name 'Aerodrome' – 'dromos' was the place where running events took place, hence the later use of the word aerodrome for an airfield.

LATHAM, HUBERT (1883-1912) – pioneer Anglo-French aviator. Latham was half English and half French, an appropriate combination as his best-known exploits involved actual or attempted flights between the two nations.

In 1904 he crossed the English Channel by balloon. Five years later he made the first attempt to make the crossing by aeroplane, on 19th July 1909. After seven miles his engine failed and he ditched his Antoinette monoplane. He was found awaiting rescue sitting on the floating wreckage smoking his obligatory cigarette.

He ordered a new Antoinette and was ready again on 25th July, but the member of his team charged with waking him early failed to do so and he awoke to the sound of Blériot departing.

He tried again two days later, but once again the fickle Antoinette engine put him into the water, this time within sight of the crowds awaiting him at Dover. His face was cut badly when ditching.

Later that year he created a sensation at a flying meeting at Blackpool when gales were keeping participants firmly on the ground. In 40 mph winds (64 kph) he opened the hangar doors and prepared for flight while all around pleaded with him not to fly.

Not wishing to disappoint the crowd, he flew for some time in those unprecedented conditions, reportedly moving watching Frenchmen to tears.

In 1910 his Antoinette engine gave him yet another ducking, this time off Nice. It seemed that flying must surely claim his life, but in the end it was a buffalo which killed him.

LEAR, WILLIAM (1902-78) – doyen of electronics and business jets. An electronics engineer, Bill Lear had developed car radios and the first successful autopilot suitable for jets.

In the early 1960s he started an ambitious venture in launching a family of business jets. The first flew in 1963, oddly enough based loosely on a Swiss design for a fighter. Over 2,000 Learjets have been sold, the company now being part of the Canadian Bombardier group.

LEDUC, RENÉ (1898-1968) – ramjet pioneer. The ramjet, or 'athodyd', is the simplest form of jet engine. In essence it is simply a shaped duct in which fuel is burned.

It is efficient at high speeds, but will not work at all until there is sufficient forward speed, so an aircraft powered by a ramjet needs another engine or a rocket for take-off.

Leduc made small ramjet units in the 1930s and exhibited models at the 1938 Paris Salon. The first full-size aircraft, the Leduc 010, flew under power in 1949, after launching from the top of an airliner.

Tests continued with improved versions until 1956. The trials were promising, but the drawback of needing another source of propulsion for take-off and ground movement was too great for operational use.

The aircraft design was unusual in that the pilot sat in a capsule projecting forward from the centre of the air intake. How would he have escaped in emergency? The whole capsule was supposed to be jettisoned, whilst deploying the parachute from stowage above the wing, or that was the theory.

LEIGH-MALLORY, TRAFFORD (1892-1944) – Fighter Command Group leader. As commander of 12 Group, covering most of the Midlands and East Anglia, Leigh-Mallory held one of the key positions in the Battle of Britain. Of the area groups, 12 was second in importance only to 11 Group, which covered the vital South-East area and handled the brunt of the attacks.

He believed German bombing raids could best be countered by building up his fighter force into 'big wings' of three or five squadrons. He bitterly criticised **Keith Park,** his opposite number in 11 Group, for attacking in squadron strength only. Regrettably, much animosity developed between the two.

Leigh-Mallory had more chance to test his tactics when he replaced Park as commander of 11 Group after the Battle of Britain, and later he became head of Fighter Command.

Towards the end of 1944 he was given a new appointment to take charge of RAF operations in the Far East. The York in which he was travelling crashed in mountains near Grenoble, France.

The Big Wing Controversy

So who was right, Park or Leigh-Mallory? When put to the test, big wings did achieve some notable successes, but there were other occasions when they took too long to assemble and the attackers had time to complete their raids. Their supporters had an answer to that one too – they claimed they were called to action too late.

Arguments and counter-arguments raged, but it is likely that the big wings were suitable to counter some threats but not others. Clearly there was much less chance to use them in Park's 11 Group near the South coast than there was for Leigh-Mallory further north.

LEVAVASSEUR, LÈON (1863-1922) – French aero-engine and aeroplane builder. A specialist in electrical engineering and petrol engines, Levavasseur decided on building a lightweight aero-engine as early as 1902, before anyone had successfully flown an aeroplane. He gave the engines, and later his aircraft, the name Antoinette after the beautiful daughter of his backer, Jules Gastambide.

In 1908 the first successful Antoinette monoplane flew, and this elegant machine was one of the outstanding pre-1914 designs. However, later Antoinettes failed to match the early promise and the company, which had contributed so much to French aviation, ceased business.

LILIENTHAL, OTTO (1848-96) –successful gliding pioneer. Lilienthal spent much of the early part of his life studying the flight of birds. In 1889 he published the result of his research, *Bird Flight as the Basis of Aviation.* Some of the data proved to be incorrect but his theoretical work and gliding trials inspired **Octave Chanute** and the **Wrights,** and this is where he formed a vital step in the story of flight.

It was also in 1889 that he started his gliding experiments. He made some 2,000 flights over the next seven years. A little problem like the lack of a suitable hill locally did not deter him – he built his own artificial one! His early gliders relied on shifting body weight for control, but later he was testing wing-warping, anticipating the Wrights who followed the same course. He also tested the now traditional aeroplane layout of an aft fin and tailplane.

By 1896 he was experimenting with rather ineffective flapping wing-tips driven by a small engine, but in August of that year he stalled in one of his gliders and was killed.

LINDBERGH, CHARLES ('Slim') (1902-74) – flew solo from New York to Paris in 1927. Having learned to fly in 1922, inauspiciously from an instructor who had lost his nerve and used every pretext to stay on the ground, Lindbergh bought a war-surplus Curtiss Jenny and joined the barn-storming circuit.

He expanded his repertoire with parachuting, his first jump nearly ending in disaster. He insisted on a stunt descent with an arrangement of two parachutes, but a shortage of parachute chord led to improvisation with grocer's twine at one point in the assembly, with near fatal results. As if that was not enough, he tried wing-walking too.

He joined the Army Air Corps for a short time, surviving an aerial collision by his first of four parachute descents. The other three happened in the next stage of his career, flying airmail services.

It was while flying airmail routes that Lindbergh learned of a $25,000 prize offered by Raymond Orteg, owner of a hotel chain, for the first flight between New York and Paris, in either direction. He also found out that several pilots were bidding for the honours, including **Fonck, Byrd,** and **Nungessor,** so speed was essential.

Charles Lindbergh – the "all-American hero" who fell from grace

He raised finance from various St. Louis businessmen, then approached a number of aircraft builders for a suitable machine without success, until the small San Diego firm of Ryan undertook to meet his requirements within 60 days.

Designer Donald Hall produced a high-wing monoplane with a single 220 hp Wright Whirlwind engine. Lindbergh had specified that the fuel tank be placed in front of his seat for safety in case of accident, but this gave him minimal forward vision.

He preferred a small, single-engine machine to the large aeroplanes with multiple crews favoured by his competitors. The Ryan was named *Spirit of St. Louis* after his backers.

The other Orteg aspirants suffered mishaps or total disaster until he alone was on Roosevelt Field (which Byrd had reserved but generously let Lindbergh use) at dawn on 20th May 1927.

Take-off, as ever on such flights at that time, was fraught, and wires at the end were cleared with little to spare. Having cleared that hurdle, he narrowly escaped losing his vital charts through the open window. Thereafter by far his greatest problem was staying awake for the next 33$^1/_2$ hours, and he had some worrying moments on that account.

He landed at Le Bourget, Paris, to a tumultuous welcome, despite it being well after dark.

His reception at Croydon a few days later, and back in America, brought crowd scenes and hero-worship almost without precedent, although he was to lose that goodwill later.

He flew to China in 1931 with his wife, Anne, in a Lockheed Sirius seaplane. They were almost drowned when being lowered into the water whilst they were already in the cockpit. He made further long-distance proving flights for Pan American Airways, but then became better known, or notorious, for his activities on the ground.

Fears for the safety of his family after the murder of his baby son led him to settle in Britain in 1935, where he stayed for four years. During this time he was invited to tour aircraft factories in Germany. The Nazis fooled him into thinking their air power was several times its true strength, leading him to believe they were invincible.

He urged other countries to settle with Germany at all costs, and after the start of the war campaigned to keep America neutral. He went so far as to include anti-Jewish and pro-German remarks in his speeches, bringing hostility on himself.

However when Japan attacked America he offered his services, and was assigned training duties in the Pacific. Against orders, he flew combat missions until he was recalled. He never lacked personal courage, for if such a well-known person had fallen into Japanese hands his fate would have been grim.

In post-war years he handled many consultancies, including work for the USAF and, once again, Pan American. He regained some of his former respect. He devoted his last ten years to wildlife conservation.

Don't Cut It!

In 1954 a film of his solo flight was made, starring James Stewart. A replica Ryan was made, and for the take-off shots the pilot was supposed to clear the telegraph wires by a small margin.

He misjudged it and struck the wires, but managed to climb away.

The directors felt it would be a pity to exclude this piece of drama, so it was retained, although not strictly correct.

LIPPISCH, PROFESSOR ALEXANDER (1894-1976) – pioneer of tail-less and delta aircraft. The long line of Lippisch tail-less designs started with a glider in 1921. He progressed through a series of Deltas (the name he used as well as the shape) in the 1930s.

His most exciting creation was his wartime tail-less, swept-wing, Me 163 rocket fighter. Its 600 mph (960 kph) speed was sensational for its day, but in service it was a failure. Its lethally explosive fuels blew several Me 163s and their pilots or ground crews to pieces, and with fuel for only a few minutes under power, even its speed was not the advantage it might have seemed.

Lippisch settled in America after the war and his work was the basis for the line of Convair jet delta fighters and B-58 bomber.

LITHGOW, MICHAEL (1920-63) – test pilot. A wartime Fleet Air Arm pilot, Lithgow took part in the Swordfish attack which crippled the *Bismarck,* although not before some of the formation had attacked *HMS Sheffield* in error. Later, whilst flying a Fairey Albacore in bad weather at night, he flew into the sea.

He and his crewman were in the water for five hours. The carrier searched for them and passed close enough in the darkness for their shouts to be heard, reflecting fine seamanship by *Formidable's* officers in deducing the correct spot and incredibly good fortune.

He became a test pilot for Supermarine, and on 25th September 1953 he set a world speed record of 737.7 mph (1180 kph) in a Swift.

He handled the first flight of the BAC One-Eleven airliner on 20th August 1963, but two months later stalling tests on the airliner led to tragedy. The aircraft entered a deep-stall, in which the high tailplane on top of the fin was shielded by the wing, and the rate of descent forced the elevator up. It was impossible to regain control and Lithgow and the six others aboard were killed.

There was criticism in test-flying circles about the company's omission to fit a tail parachute, which could have avoided the accident.

Testing Tribulations

The One-Eleven accident showed that test flying still had its risks, even if they are lower than they once were. There were to be two more mishaps before the aircraft entered service.

A tail parachute was later fitted and on one of the incidents the pilot deployed it but then forgot to release it, leading to a forced landing in a clear area of Salisbury Plain.

LOBELLE, MARCEL (1893-1967) – Chief Designer for Faireys. Lobelle designed the Fairey Fox light bomber in 1924, a sensation in its day when it outpaced contemporary fighters.

By far his most creation was the Swordfish biplane torpedo bomber. Despite being obsolete by the beginning of the war, it was one of the few pre-war types still in service at the end. It even outlived its planned successor, the Albacore.

Lobelle remained at Fairey until 1940. He set up his own company, ML Aviation, later renamed ML Engineering. It made ancillary equipment but he was not quite finished with complete aircraft. His last design was an inflatable aircraft in the 1950s. It was intended for military use, to be carried deflated in a vehicle, inflated and flown away. Prototypes were flown but it was never put into production.

The Stringbag – The Unlikely Success Story

The Swordfish, widely known as the 'Stringbag', was slow and apparently vulnerable, but courageous crews achieved astonishing success in it. Much of the Italian fleet was sunk at Taranto with remarkably light losses, the battleship Bismark was disabled by a Swordfish torpedo, and in another attack four ships were sunk with three torpedoes! It was surprisingly agile – one pilot evaded two Italian fighters until his pursuers flew into the sea. However, inevitably there were other occasions when there were heavy losses.

LOUGHEAD, ALLAN (1889-1969) – founder of the Lockheed company. Loughead is reputed to have made his first flight (solo!) in a Curtiss biplane in 1910. The story is that no-one else could persuade it to leave the ground that day, until he climbed in and away he flew.

Whether or not it was quite like that, in the following year he started designing his own aircraft with his brother, Malcolm. It flew well and the brothers were able to keep themselves with joyriding income. They still needed to engage in other fields to prosper and Malcolm started the Lockheed brake business, used on cars ever since. Because of widespread mispronounciation of their name, they traded under the simpler spelling of Lockheed from 1926.

Their aviation business lurched from one financial crisis to another, but in 1927 came salvation with their highly efficient high-wing Vega monoplane, designed by **Jack Northrop**. The Vega set innumerable records, including two round-the-world flights by one-eyed **Wiley Post.**

In 1929 the company was taken over and Allan Loughead left. The new parent company went bankrupt in 1931 but was saved by the narrowest of margins and became one of the 'giants' of aviation. Loughead tried once more to set up in aircraft building but built just one prototype.

LOVELL, JAMES (1928-) – astronaut, commander of *Apollo 13.* Lovell became one of the most experienced of all astronauts. He flew on the *Gemini VII* mission, and was one of the three crew on the first flight around the moon, in *Apollo 8.*

His most outstanding moment came on a failed mission. *Apollo 13* was to have been the third moon landing. Whilst en route to the moon Lovell reported "Hey Houston! We have a problem here." It was a classic understatement – a major explosion had crippled the spacecraft and their chances of survival looked grim.

They still had to round the moon, for the spacecraft could not return without using the moon's gravity to assist its trajectory. Power and oxygen had to be severely rationed, and to this end the three crew transferred to the lunar module, using it as a 'lifeboat'.

Millions around the world listened to their progress, prayed, and rejoiced at their safe return. Lovell's calm handling of a situation which could well have left them stranded in space was universally admired.

LYNCH, BERNARD (1918-) – made first recorded ejection from an aircraft. When Martin-Baker were developing their earliest ejection seat, Lynch, a fitter, elected to ride the test rig. After many simulated ejections, the time came to test the device in an aircraft. Several ejections were made using dummies, which was just as well as there were failures, before the first live test.

Lynch made his historic exit from a Meteor, flown at 320 mph (515 kph) and 8,000 feet (2,440 m) on 24th July 1946. He made further test ejections, but paid for his bravery with damage to his back. He was awarded a well-deserved British Empire Medal.

MACCHI, GIULIO – Italian aircraft constructor. Macchi started in 1912 by building Nieuports under licence, but quickly started on his own designs. From 1915 the company became renowned for its seaplanes. In 1921 a Macchi M7-bis won the Schneider Trophy, and in 1934 his MC72, designed by Mario Castoldi, set a world speed record of 440 mph (708 kph).

The Macchi MC202 of 1941 was probably the best Italian wartime fighter. The company still exists, trading as AerMacchi, specialising in jet trainers.

MacCREADY, DR. PAUL (1925-) – man-powered aircraft designer. A leading American glider pilot, MacCready turned to the problem of man-powered flight in the 1970s. Many had tried for an elusive prize offered by an industrialist, Henry Kremer, for a man-powered figure-of-eight Some had flown straight and level, but none had managed the necessary turns.

MacCready studied aerofoils suited to speeds of around 5 mph (8 kph), not a requirement often met by other designers! His *Gossamer Condor* used an ultra-light structure covered in Mylar film, giving a weight of just 70 lb (31 kg) despite its 96 ft (29 m) span. It needed one-third of a horse-power for level flight. Athlete **Bryan Allen** took the Kremer prize at last in 1977.

Two years later he built the *Gossamer Albatross*, a new design made with carbon-reinforced plastics and polystyrene foam. In June 1979 Bryan Allen made the first man-powered crossing of the English Channel, winning a further prize given by the ever-generous Kremer.

In 1980 he built a solar-powered aircraft, the *Solar Challenger.* On 7th July 1981 it was flown across the English Channel by Steve Ptacek.

MACH, ERNST (1838-1916) – devised the Mach scale for measuring speeds. Austrian born Mach proposed the scale in 1887. Needless to say he was not concerned with manned supersonic flight! He was in fact studying projectiles. Speeds are measured in multiples of the speed of sound.

Mach 1 is approximately 761 mph (1,220 kph) at sea level and 15°, falling to around 660 mph (1,060 kph) by the stratosphere. The scale was named after him when aircraft speeds started to approach that of sound.

MALAN, ADOLPH ('Sailor') (1910-64) – leading RAF fighter pilot. South African-born Malan joined the RAF in 1935 after serving as an officer on merchant ships, hence his popular name.

Malan was credited with 32 victories during 1940 and 1941, including some of the first at night in single-seat fighters. He commanded a squadron in the latter part of the Battle of Britain and later became a wing leader before promotion to Station Commander at Biggin Hill in 1943.

He used his experience to draw up his *Ten Rules of Air Fighting,* giving valuable guidance on how to win over enemy pilots and, probably uppermost in most men's minds, how to stay alive.

He returned to South Africa in 1946 and took up farming.

They Flew for Freedom

Lack of space prevents inclusion of even many of the best-known Battle of Britain pilots, and in a sense it would be unfair to do so, for thousands of aircrew flew in little-known but heroic actions which never received the (rightful) acclaim of that great battle.

'Sailor' Malan is included as representing the many who came from homes far from Britain to serve in the RAF. Many would never see their homelands again.

MANNOCK, EDWARD ('Mick') (1887-1918) – highest-scoring British pilot of World War One. Credited with 74 victories, including 24 in one month, Mannock favoured team tactics, unlike some of the earlier pilots such as Albert Ball who flew alone. Times and tactics were changing.

Mannock took care in building up confidence of inexperienced pilots, and it has been claimed he credited some with victories which were properly his own. If so, it is marked contrast to some other claimants of high scores! Remarkably, he was a fine shot despite poor vision in one eye.

Almost inevitably, his 'number came up' in the end, in his case probably due to ground fire, in July 1918. His VC was awarded after the war as a result of lobbying from colleagues that he had been overlooked for the honour.

MARTIN, GLENN (1886-1955) – founder of the Glenn L Martin company. Inspired by the work of **Octave Chanute** and **Samuel Langley,** Martin started experiments with gliders in 1905, progressing to a powered aeroplane in 1909, in which he taught himself to fly. He formed his first company in that year, but following a merger reformed it as a new concern in 1917.

Martin tended to concentrate on flying-boats and bombers. Notable designs were the M-130 airliner flying-boats of the 1930s, the wartime Mariner flying-boats, Maryland and Marauder bombers. Over 5,000 of this last-named were built, including some for the RAF.

In post-war years he built Canberra bombers under licence as the B-57, his modestly-selling Martin 202 and 404 airliners, and a rare jet flying-boat, the Seamaster. Soon after his retirement in 1953 the firm concentrated on missiles and space equipment. In 1994 the Martin company merged with Lockheed.

MARTIN, SIR JAMES (1893-1981) – ejection seat pioneer. The Martin-Baker company was formed in 1934. The co-founder was Valentine Baker, a former flying instructor.

Martin designed a range of capable fighters, but none were ordered into production. In August 1942, Baker, his partner, was killed landing the MB-3 after engine failure. Martin's MB-5, first flown in 1944, was probably one of the finest piston-engine fighters ever built, but by then the jet age had nearly arrived and only the prototype flew.

However it was the coming of jets which gave Martin his great achievement in life. Climbing out of the cockpit in emergency was clearly too slow at jet speeds and some better means of escape was becoming vital. After toying with spring-loaded booms mounted along the fuselage, Martin turned to the idea of the ejection seat. The first tests with dummies were carried out in 1945 from a Defiant, leading to **Bernard Lynch's** first live test on 24th July 1946. The first life to be saved was that of Armstrong Whitworth test pilot John Lancaster, who ejected from an AW52 'flying wing' on 30th May 1949.

Early 'bang seats' could not be used at low speeds or altitudes. Progressively they were improved, J Fifield making the first ground level test on 3rd September 1955, followed by 'Doddy' Hay making the first 'zero-zero' (ground level, no forward speed) test on 1st April 1961.

Since then rocket seats and devices to assist under-water escapes have been added. The seats are used all over the world, and some 6,500 aircrew have offered their 'thanks to James Martin'.

MASEFIELD, SIR PETER (1914-) – leading personality in British civil aviation. In a remarkably wide career, Sir Peter started with Faireys then progressed through aviation journalism, a government advisor, Air AttachÈ to the USA, then in turn heads of BEA, Bristol Aircraft and British Airports Authority. Indeed there are more appointments than can be listed here.

An interesting work of research was his book on the R-101 airship disaster *To Ride the Storm.* It completely overturned the previous view that the planning, design and building of the airship had been a trail of incompetence. He showed that those involved had worked professionally, and as is true of most air accidents, the disaster was caused by an unlucky sequence of events.

When Chairman of BEA, Sir Peter was asked why they struggled to make profits whereas pre-war Imperial Airways were profitable with their lumbering biplanes. His response "That's easy, they were so slow all the money was made in the bar!"

MAXIM, SIR HIRAM (1840-1916) – tested large biplane in the 1890s. An American with a string of inventions to his credit, including the machine-gun which bore his name, Maxim performed his flying experiments in Britain, where he had settled in 1881.

Maxim evidently did not believe in working on a small scale, for his colossus with its two 180 hp steam engines of his own design weighed 3Ω tons. He tested it on a track with restraining rails, but on one trial it rose enough to foul the structure and was badly damaged. Impressive it might have been, but there was no means of control and Maxim was far from having a practicable aeroplane. Nevertheless he understood aerodynamic principles and it was a step towards flight. Maxim became a British citizen in 1900 and received a knighthood the following year.

McCUDDEN, JAMES (1895-1918) – one of the top-scoring British pilots of the First World War. Initially a soldier with the Royal Engineers, McCudden joined the RFC in 1913 as an engine fitter. His applications for flying training were rejected until 1915, on the grounds that there was a shortage of good engineers! Once airborne he quickly showed his ability and courage, eventually accounting for 57 German aircraft. He was meticulous in the standards he set for pilot training, team tactics, and equipment maintenance.

McCudden was awarded a VC in April 1918. Three months later his SE5A was seen to bank steeply after take-off and hit the ground. It was never established whether the crash was due to an aircraft fault, pilot incapacitation, or unwisely turning back after engine failure.

McDONNELL, JAMES (1899-1980) – founder of the McDonnell Aircraft Company. After employment with several aircraft companies, McDonnell started his own business in 1939. During the war years his firm built other companies' designs under licence, although one indigenous fighter prototype was made. Success came with the FD-1 Phantom naval jet fighter of 1945 which served well.

He branched out into helicopters, but despite, or because of, their technically advanced features, none went into service. Much later the company would return to rotary wings.

The aircraft which made McDonnell one of the giants of industry was the 1958 F-4, for which he revived the name Phantom. This large twin-engine fighter was designed for carrier use but became widely used by the USAF and many other air forces, including the RAF (and for a time the Royal Navy). It was the most important combat aircraft in Vietnam. Over 5,000 were made.

McDonnell saw his company merge with Douglas in 1967 to form McDonnell-Douglas. Now he was the key figure in one of the largest producers of both civil and military aircraft in the world. He died just a month after retiring in 1980.

MESSERSCHMITT, PROF. WILLY (1898-1978) – leading producer of German military aircraft. Best known as the originator of the principal German fighter in the Second World War, Messerschmitt started his career building a glider from plans he was given whilst still a schoolboy.

He progressed through a series of sailplanes, later expanding by buying the Bayerische Flugzeugwerke (Bavarian Aircraft Factory). A series of airliners followed, but a record of structural failures led to their withdrawal and he turned his attention to military production.

Messerschmitt's best-known type, the Bf109 flew in 1935 (Pre-1938 designs were prefixed 'Bf' after the works name, but then the company was renamed and thereafter they carried 'Me' numbers. This was realised in Britain only after the war, so to Battle of Britain pilots they were 'Me109s'). Its high performance and good handling attracted Luftwaffe orders.

One area he misjudged was the armament, for he provided only two machine-guns. Experience in the Spanish civil war showed the need for more, and it took considerable ingenuity to adapt the design, but the task was done and by 1940 the Bf109 was a formidable opponent to the RAF. It had roughly the speed of a Spitfire, but had a wider turning circle and was not as strong, limiting hard pull-outs from dives. Over 30,000 were made and it remained a front-line fighter throughout the war.

His Bf110 heavily armed twin-engine fighter proved disappointing in daytime operations, lacking the agility of single-engine types, but later served as an effective nightfighter.

Later came his most advanced fighter, the Me262 jet, which entered service in July 1944, within a few days of the Meteor in Britain. With its swept wings it was more advanced, but its effect was muted by undeveloped engines, slow production, and fuel shortages. In parallel, Messerschmitt built the Me163 rocket fighter, but this was an inherited design and not his own.

Not all Messerschmitt's designs were inspired. His Me210 fighter was a failure, so much so that he was removed from control of his own company, although he continued design work. Likewise his huge Me321 glider was an impractical monstrosity, and its powered derivative the Me323 was a lumbering sitting-duck target for Allied fighters.

After the war Messerschmitt restarted in Spain, until aircraft building was once more allowed in Germany. He once again became a major player in his country's aviation industry. His company merged with others to form MBB in 1969.

His working life extended to early studies on the Airbus and Tornado, where he sometimes found himself sitting alongside men whose greatest ambition had once been to blast his creations out of the sky.

Whilst establishing himself in Spain, Messerschmitt must have been struck by an irony in seeing Spanish-built versions of his Bf109, named Buchons – they were powered by the very engine which propelled its wartime opponents, the Rolls-Royce Merlin.

MIKOYAN, ARTEM IVANOVICH (1905-70) – Russian jet combat aircraft designer. The MiG company was formed in 1939 as a joint venture between Mikoyan and Mikhail Gurevich. Prior to this, both had worked for the Polikarpov design bureau (the usual terminology in Soviet aviation)

The MiG-1 and MiG-3 piston-engine fighters were made in respectable numbers, but the partners really made their name with jet aircraft. After a dabble with a rocket fighter, they turned to jets with the straight-wing MiG-9.

The spectacular advance came with the swept-wing MiG-15, which appeared in 1948. Mikoyan had significant help from captured German data on swept wings, and from Britain for the engine – incredibly, the British government had supplied a batch of Rolls-Royce Nenes and Derwents to Russia, and a straight copy of the former went into the MiG-15. Over Korea the fighter gave British and American pilots some serious opposition.

The MiG company has remained the leading Russian supplier of high-performance fighters.

MILCH, ERHARD (1892-1972) – responsible for German aircraft production and Luftwaffe expansion. Milch was an observer in World War One, but must have displayed leadership qualities, for he became commander of a fighter squadron although not a pilot (bizarre though this sounds, there have been British squadrons too led by observers rather than pilots).

In 1920 he held appointments in some of the small, usually short-lived airlines of the time, before becoming head of the newly-formed Lufthansa.

Once the build-up of the Luftwaffe began, Milch took much of the responsibility for selecting aircraft types and organising production. In so doing he managed to quarrel with most of the leading people in the industry. He was especially ruthless to **Junkers,** then an old man, whom he arrested. Junkers had helped Milch earlier in his career; presumably gratitude was not part of Milch's character.

He was second only to Goering in the Luftwaffe hierarchy, enjoying the trapping such as his private train. Eventually he fell from grace by opposing Hitler's wish to use the Me262 jet fighter as a bomber, and he was forced to resign most of his posts.

Milch was sentenced to 'life' at Nuremberg, but was released in 1954. He lived till 1972, becoming one of the last of the top Nazis.

MILES, FRED (1903-76) – builder of British light aircraft. Miles started light aircraft design in the late 1920s, and after a couple of false starts built the 'Southern Martlet' in 1929 in conjunction with the Southern Aero Club, of which he was a founder member.

Early Miles designs were built by Parnall, near Bristol, or by Phillip and Powis of Reading. Later they were sold under the Miles name and achieved a fine reputation for racing, touring, and as trainers. The company was a family business, with brother George and Fred's wife Blossom.

During the war thousands of pilots learned to fly in Miles Magisters, followed by Masters for advanced training. The Messenger communications aircraft was used by General Montgomery, among others.

Although Miles are linked with light aircraft, they designed large airliners and actually partly built what was to have been Britain's first supersonic aircraft, the M52, before the government caught cold feet and cancelled it. Times became tight after the war and the brothers even took out a licence to make *Biro* pens, but to no avail. Problems with the severe winter of 1947 were the last straw and Miles closed.

Fred and George tried again in 1957 with a neat jet trainer, the Student. It performed well but they were unable to secure orders. The new Miles company was absorbed into Beagle, the creation which was supposed to put new zest into British light aircraft (it didn't).

British Airways captain Eric Moody was flying his Boeing 747 near Java at night when all four engines ran down and failed, accompanied by loud bangs, smoke in the cabin, and strange flashes of light. They descended 24,000 feet without power, mystified as to what was wrong.

The captain's announcement to his passengers has to be the most remarkable in airline history "We have a small problem. All four engines have failed. We are doing our damnedest to get them going again. I trust you are not in too much distress."

The multiple failure was later found to have been caused by a volcanic cloud. The handling of the incident was a magnificent feat of airmanship.

MITCHELL, REGINALD (1895-1937) – designer of the Spitfire. Mitchell was involved in Supermarine designs from 1916 onwards. By the early 1920s he was wholly responsible for their designs. In a taste of what was to come, his 1922 Sea Lion took the Schneider Trophy that year.

His Southampton flying-boat first flew in 1925 and gave 10 years of exemplary service. It would be followed by other fine flying-boats, including his little amphibious Walrus, or 'Shagbat', which saved thousands of men during the war when used for air/sea rescue duties.

From the mid-1920s came his historic line of high-speed monoplane racing seaplanes for the Schneider Trophy contests. His S.4 in 1925 set a world speed record of 226.7 mph (365 kph), but was destroyed during practice for the race.

Reginald Mitchell – creator of the Spitfire

The 1927 contest was won by his S.5, and in 1929 it was the turn of the S.6, with Rolls-Royce 'R' engine replacing the Napier Lion of the S.5, to take the trophy. Soon afterwards it too set a world speed record, this time at 357.7 mph (564 kph). The comparison with the record of just four years earlier shows the advance made under the stimulus of these races.

In 1931 Mitchell produced a yet further improved version, the S.6B, aided by more power from the R engine. As it happened the other competitors had to withdraw and the Schneider Trophy was taken without opposition. Mitchell believed that the improvements in engine powers, airframes and fuels would have taken three times as long without the pressure of the race.

Mitchell built a monoplane fighter with cranked wing and fixed undercarriage. It was named Spitfire. Its performance was poor and he started on a new design. The company chairman, Sir Robert McClean, firmly told the Air Ministry that "No official interference with its design would be tolerated."

Mitchell was suffering from cancer, and when in Germany recovering from treatment his meetings with the aviation community convinced him that war was to come. He resumed work on his new fighter with renewed frenzy.

First flight of the Spitfire was on March 5th or 6th (records were lost in a bombing raid) 1936, in the hands of **'Mutt' Summers.**

Mitchell saw his greatest creation take to the air, but never saw it in service or knew the part it played in saving his country. The cancer returned and he died in 1937. Spitfire development continued under his successor, Joseph Smith, whose part in keeping the fighter abreast of its opposition throughout the war years is often overlooked.

Spitfire

The Spitfire captured the imagination of the British public in a way unrivalled by any other aircraft. Its graceful elliptical wing made it easily recognisable and the name 'Spitfire' added to its aura. Even German pilots were afflicted, for some were adamant they had been shot down by Spitfires when records showed that other types had to be responsible.

The originator of the name is unknown. One man who disliked the name was Mitchell, possibly because of its use for his earlier fighter.

Over 22,000 Spitfires were built, not without difficulty for the wing was complex to make. It remained in production and front-line service throughout the war.

Today an appearance by a Spitfire at any airshow will make it a star turn, aided by the stirring sound of the Merlin engine.

MITCHELL, WILLIAM ('Billy') (1879-1936) – leading exponent of American air power. As soon as he learned to fly in 1916, Mitchell was campaigning for a separate air arm to the US Army. He flew in combat in the First World War, and met **Trenchard.**

He was influenced by the British commander's ideas and back in America he sought to make the nation more aware of the need to strengthen its air power.

In 1921 he mounted a demonstration to convince the US Navy of the vulnerability of its ships by bombing and sinking some old warships. The admirals pointed out, with some justification, that the ships were anchored and not firing back, but there were valid lessons and if they had been heeded there might have been no Pearl Harbour debacle.

Always controversial, he was court-martialed in 1926 after outspoken comments about both Army and Navy policies. Nevertheless he is credited with playing a major part in the establishment of the US Army Air Corps, later Air Force, and in building American air strength.

His name was commemorated in the B-25 Mitchell bomber, and most unusually, in 1947 he was awarded a posthumous elevation in rank to major-general.

MOELDERS, WERNER ('Vati') (1913-41) – leading Luftwaffe pilot and tactician. As a pilot with the Legion Kondor in Spain, Moelders flew in combat for the first time and devised the Schwarm formation, or *finger-four* as it was later called in the RAF. Four aircraft flew in two pairs in loose formation, roughly in the shape suggested by the English name. The fighters were at different heights, giving good all-round views of possible attackers and freedom for action. It was some time before the RAF adopted it to replace their vics of three, impressive at air displays but less so when used 'for real'.

In his combat career he logged 115 victories, including his 14 in Spain. Many were on the Russian front. In 1940 he was captured in France but released after the country surrendered.

He became Inspector General of Fighters in 1941, but shortly afterwards was killed in a flying accident on the way to the funeral of his friend **Ernst Udet.**

MOLLISON, JAMES (1905-59) – long-distance trailblazer. Mollison started his quest for records with a solo flight from Australia to Britain in 1931. He followed it with a record to Cape Town in a Puss Moth. On arrival he met the woman pilot **Amy Johnson,** whom he was to marry a few months later. Each of the successful flights was preceded by a failed attempt in which he wrecked his aeroplane. Sponsors needed deep pockets and plenty of tolerance.

In August 1932 Mollison made the first East-West solo North Atlantic crossing, followed by a solo South Atlantic crossing the next year. In July 1933 the husband and wife team made a transatlantic flight in a de Havilland Dragon, intending to set a distance record on their return. They reached Connecticut but crashed on landing, probably due to tiredness, one of the greatest dangers on such flights. In the following year they entered a Comet for the MacRobertson Race to Australia, but retired with technical problems.

It was no secret that the Mollisons had heated rows and they duly divorced. They may have been too similar in flying ability. Some pilots found it better to have someone with much less experience to accompany them, so leaving decisions of captaincy in one pair of hands.

Jim Mollison flew with the transatlantic ferry service in World War Two.

MONTGOLFIER, JOSEPH (1740-1810) and Etienne (1745-99) – the brothers who first launched mankind into the skies. The Montgolfiers were papermakers who started experiments with small hot-air balloons in November 1782. There are various stories about how they formed the idea, but Joseph attributed his interest to a book by the scientist J. Priestley. They believed combustion created a light gas and never realised it was heating the air which caused it to rise.

In April 1783 a larger 35 ft (11 m) diameter balloon was released. It landed three-quarters of a mile (1.2 km) away. A series of balloons was tested, until an ascent before the king and queen of France was scheduled for 19th September. A week beforehand, the balloon was wrecked by a storm. By enormous effort and 24-hour shift work, another balloon, 57 ft high (17 m), was completed in time. A cargo of livestock, comprising a sheep, a cock and a duck was put aboard and with much ceremony and booming of cannons the balloon was released into its element. Thus a humble sheep became the first creature to fly not intended by nature to do so. The balloon landed two miles away 3.2 km). The prophets of doom pointed to a damaged wing on the cock and cited it as a hazard of air travel, but others had seen the sheep kick the cock before launch.

After this success, a man-carrying balloon was started. It was about 75 ft high (23 m). A gallery was built around its neck, and openings were provided to enable the crew to add fuel to the fire basket suspended below. **Pilâtre de Rozier,** nominated to make the historic first flight, practised with captive ascents.

Mankind's first true flight came on 21st November 1783, with the **Marquis d'Arlandes** as the second man aboard. The flight is described under their entries.

The 'Montgolfière' hot-air balloons were soon eclipsed by those using gas, but the brothers had led the way in showing that flight was possible, and that no harm would come to the aeronauts – for some reason many had believed that the lower pressure at height would be harmful, despite the experience of mountaineers who had been far higher. Even in ballooning, their hot-air type would make a come-back, for the vast majority of modern balloons use their principle.

Keeping their Distance

The Montgolfiers, incorrectly deducing why their fire produced lift, believed the substances burned were significant and used a concoction of straw, wool, old shoes and decomposed meat. Not surprisingly, when the king and queen came to inspect the balloon they were 'obliged to retire' to a more congenial distance.

MOORE-BRABAZON, JOHN THEODORE CUTHBERT (Lord Brabazon of Tara) (1884-1964) – first British national to fly in Britain. A versatile sportsman who won honours in motor-racing, bobsleighs, sailing and ballooning, Moore-Brabazon turned to aeroplanes in France in 1908. The flight which put him in British record-books was on 30th April 1909, in a Voisin biplane.

Moore-Brabazon *(continued)*

Later that year he won a £1,000 *Daily Mail* prize for a circuit in a British aircraft, a Short biplane (licence-built Wright).

He applied to join the RFC but was assigned technical work rather than flying duties. He played a major role in the use of radio for aviation and in using aerial photography.

In the inter-war years he became an MP, and in 1942 was appointed Minister of Aircraft Production. He held the post for a short time only, reputedly being moved to the Lords following an injudicious remark. In 1943 he headed the Brabazon Committee to plan post-war transport aircraft needs. Some of the ideas became successes, but the great airliner which bore his name was conceived without examining the market and scrapped unwanted by airlines.

He kept up aviation connections and lived life to the full right to the end, as when he descended the Cresta Run at 70 years of age.

MOOREHOUSE, WILLIAM (?-1915) – first air Victoria Cross. The award followed a solo raid on troop concentrations on 26th April, 1915. The recognition was posthumous, for he died of wounds sustained in the attack.

Heroes Unsung

There is insufficient space to include every holder of even this supreme gallantry award. It should be appreciated, too, that for every VC there were many others whose deeds were not witnessed or were simply of lower rank than the one so recognised. There were untold numbers of heroes without medals.

MOZHAISKI, ALEXANDER (1825-1890) – Russian bid for the first aeroplane flight. Mozhaiski just became airborne when his big monoplane trundled down a ramp in 1884. His machine was powered by two steam engines, so he can claim to have made the first twin-engine aircraft, even if it was a failure.

Much later, Russian propaganda made exaggerated claims about his trials, suggesting that he beat the Wrights into the air.

NESTEROFF (or 'Nesterov') PETER NICHOLAIVICH (1887-1914) – first to fly a loop. Nesteroff is generally credited with the first loop, at Kiev on September 9th 1913 in a Nieuport, although 'credited' was not a term his Russian army superiors used. He was disciplined and placed under house arrest for endangering government property.
He was killed in combat in the First World War. Reportedly he rammed an enemy aircraft. The town near where he was killed was renamed after him.

NEWMAN, LARRY (1947-) – first transatlantic balloon flight. The crossing in Double Eagle is described under **Ben Abruzzo.**

DE NIÉPORT, EDOUARD (1875-1911) and **CHARLES** (?-1913) – early French aircraft and engine designers. Edouard was the brother who started the interest in flying. Having designed magnetos and other electrical items for cars and aeroplanes, he built and flew his first machine in 1908.

The two brothers formed the Nieuport company in 1909. Edouard had slightly altered his name to disguise his participation in cycle racing from his family lest they thought it hazardous.

What they thought of him flying is not recorded, but if they viewed it with foreboding they did so with justification - Edouard was killed in a crash in 1911, followed by Charles two years later.
The Nieuport company built some of the best Allied aircraft of World War One. The firm continued in business until nationalised in the 1930s.

NOORDUYN, ROBERT (1893-1959) - Dutch-born builder of 'bush' aircraft. Noorduyn went to North America to set up production there for Fokker and formed his own company in Canada in 1934. He designed just one aircraft which entered production, but so tough and versatile did it prove that although it first flew in 1935 a number are still flying 60 years later.

The Norseman was a rugged high-wing aircraft built to take wheels, floats or skis, and especially intended for Canadian bush flying. The survivors of the 903 built are to be found, not carefully cosseted as preserved exhibits, but still hard at work doing the job they were built to do. There are few aircraft types which have proved so durable. Noorduyn sold his company in 1946.

NORTHCLIFFE, LORD (Alfred Harmsworth) (1865-1922) – stimulated early aviation with his prizes. As proprietor of the *Daily Mail*, Northcliffe was keen to promote aviation and offered a series of prizes for milestones in flight. Notable among these were the prizes for the first aeroplane crossing of the English Channel, won in 1909 by **Blériot**, and the first non-stop Atlantic flight, gained by **Alcock** and **Brown** in 1919.

His good work was continued by his brother, Harold Harmsworth, later Lord Rothermere, who continued the campaign to strengthen British aviation in the 1930s.

NORTHROP, JOHN ('Jack') (1895-1981) - founder of Northrop Aircraft, pioneer of multi-spar wings and of tail-less aircraft. Northrop alternated between Lockheed and Douglas from 1916 to 1928, when he formed his own company, which was still linked with Douglas. Among his designs was the very efficient Lockheed Vega which set new standards in performance for its day.

Perhaps of even greater importance was his work on wing design. He developed the concept of stressed-skin wings with multiple spars which became universally adopted, although it took a surprisingly long time for some manufacturers to follow his lead. His triple-spar wings gave the Douglas airliners much of their legendary durability - some have been flying for nearly sixty years and their wings never break.

Another of his enthusiasms was the 'flying wing'. By dispensing with a tail and much of the fuselage, there should be a large reduction in drag. The problem faced by designers before the age of computer controls lay in fore-and-aft stability. Northrop flew such a machine in 1928 and was to return to the idea later.

In 1942 he produced the world's first dedicated night-fighter, the twin-boom P-61 Black Widow. It scored a good combat record.

Northrop tried again with flying wings, developing the B-35 four-engine bomber, but its complex propeller drives gave untold trouble. At that time, in the mid-1940s, he tested a wide variety of projects, including a rocket aircraft and a remote controlled flying bomb. Most sensational of all was his mighty B-49 eight-engine jet bomber flying wing, flown in 1947. It would look futuristic even today, but there were problems with its handling and a fatal accident to one prototype probably swung the decision to cancel production.

One of Jack Northrop's more bizarre ideas was his F-79 Flying Ram jet fighter of 1945. It was designed to ram enemy aircraft, probably the only example in history planned to do so. It was claimed it could ram ten enemy per sortie and survive the impacts (but could the pilot?). In the event, the prototype crashed and the idea was dropped, no doubt to the relief of pilots who might have put it to the test.

NUNGESSER, CHARLES (1892-1927) - French fighter pilot and transatlantic flight aspirant. Nungesser started flying before the First World War, reputedly making his first flight solo, jumping into an aircraft and flying it after being refused a flight. He became one of the leading French combat pilots, with 45 victories, but was himself badly wounded several times. As if his flying injuries were not enough, he added more from a car accident.

He attempted a transatlantic flight from east to west in 1927 with Capitaine Coli. They took off in *l'Oiseau Blanc* from le Havre on 8th May. Reports of their being seen over Newfoundland and along the Canadian seaboard trickled in and radio networks followed their 'progress'. Alas, every one of the sightings was false. The two airmen were never seen again after leaving France.

OHAIN, HANS VON (1911-) – designer of the first jet engine to fly. Whilst a student at Gottingen University, von Ohain patented a jet engine in 1935. It featured a centrifugal compressor and (rather inefficient) radial inflow turbine.

Supported by **Heinkel**, he ran an engine in 1937, using hydrogen as fuel. It reached 550 lb (250 kg) thrust. By 1939 he was running his HeS 3B unit, which delivered 1,100 lb (500 kg) thrust, using liquid fuel. One such unit, in the specially built He178 airframe, became the first jet to fly. Flown by Erich Warsitz it made a brief hop on 24th August 1939, and a full flight on the 27th. This latter date is generally considered to mark the first jet flight.

Two improved units were fitted in the Heinkel 280 fighter in 1940, which reached around 500 mph (800 kph), but progress then became bogged down by bureaucratic infighting and it was not until 1944 that a German jet fighter, the Me 262, entered service.

He continued work on jet engines throughout the war, tackling the more demanding axial compressor layout. The German engines which saw service were more technically advanced than British units but were less developed and gave poor reliability.

PARK, SIR KEITH (1892-1975) - Group commander during the Battle of Britain. A New Zealander, Park was a soldier in the early part of his military career, was wounded on the Somme and fought at Gallipoli. He transferred to the RFC and shot down 20 enemy aircraft.

Park, Sir Keith *(continued)*

In 1940 he commanded 11 Group, covering South-East England and the area which took the brunt of attacks. Most historians now credit Park and his superior, **Dowding,** with masterful tactical handling of the battle. They made the best use of scarce resources, taking care whenever possible to rest squadrons before pilots began to 'crack' under pressure. He spent much of his time visiting units in his Hurricane, discussing their problems and tapping their experience.

Regrettably he had not only Germans to fight. His opposite number in 12 Group, **Leigh-Mallory**, attacked him incessantly for his tactics and forever putting the case for his 'big wings'.

Park and Dowding were displaced from their posts towards the end of 1941. Many have seen this as an injustice, to some degree justifiably, but Park had been under great strain and was due for a rest.

He held further appointments in Training Command, Air Defence, in Malta, the Middle and Far East. He returned to New Zealand in 1948.

PARKE, WILFRID (?-1912) – first known recovery from a spin. Parke was approaching to land in an Avro biplane in 1912 when he accidentally stalled, reportedly at 600 ft (200m) and entered a spin. If this height is correct he was lucky to have survived! He tried various control movements without success until he applied opposite rudder to the direction of the spin, whereupon he recovered, it was said at 50 ft (15 m). Where he made a real contribution to flying was in analysing what had happened and setting out the recovery routine. For a time, a spin was known as 'Parke's Dive'.

He was killed flying a Handley Page monoplane in December of that year. Ironically, having set out the rules for recovering from one situation, be broke what became another golden rule, of not turning back following engine failure just after take-off.

PATTLE, MARMADUKE THOMAS ST. JOHN ('Pat') (1914-41) – outstanding wartime RAF combat pilot. It is possible that 'Pat' Pattle shot down as many as 50 enemy aircraft, but as so many were over water some could not be confirmed. South African-born Pattle scored most of his victories in the Middle East area, a remarkable number whilst flying obsolete Gladiator biplanes. He was himself shot down while heavily outnumbered.

PAULHAN, LOUIS (1883-1963) – French pioneer aviator. The tenth Frenchman to learn to fly, by the end of 1909 Paulhan was one of the leading pilots at air 'meets'.

In 1910 he took several prizes and set an altitude record at a competition in Los Angeles. A planned tour of the United States was curtailed when he was grounded by **Wilbur Wright** taking action for supposed patent infringement.

His best-known feat was winning a £10,000 prize offered by **Lord Northcliffe** for the first flight between London and Manchester. It so happened that both Paulhan and **Claude Graham-White** were ready to bid for the prize on the same day, April 27th 1910, and both were flying Farmans, so a dramatic race was in prospect. The press made the most of the drama and there was frenzied public interest. After an overnight stop, Graham-White took the lead by taking off before dawn. Paulhan stormed after him and overtook him when his rival ran into technical trouble. Paulhan landed in Manchester to be welcomed by a crowd of reportedly 100,000.

Paulhan turned to design and manufacturing during the war to come, then left the aviation scene. He was to fly once more from London to Manchester, to mark the 50th anniversary of his triumph. This time it was quicker – 48 minutes in the back seat of a Meteor.

The London to Manchester race was exciting but hardly demonstrated the potential speed of air travel – Paulhan's family and friends kept up with him in a special train!

PEARSE, RICHARD (1877-1953) – New Zealand pioneer. Claims have been made that Pearse flew in 1903 before the Wrights. He stated, however, that his attempts were in 1904, and that they were unsuccessful. In view of his own denials, it is strange that the claims persisted so long, even to the extent of being perpetuated in a television documentary.

PEGG, ARTHUR JOHN ('Bill') (1906-78) – Chief Test Pilot for Bristols. Pegg became an RAF pilot in 1925 and transferred to test flying at Martlesham Heath in 1930. Whilst diving an Avro Tutor biplane trainer, the upper wings came off at 3,000 ft (930 m) and he was lucky to be thrown out.

He joined Bristols in 1935 and tested all the company's wartime products, experiencing the usual crop of incidents typical of the test pilot's life at that time.

He became Chief Test Pilot in 1947. On 4th September 1949 he flew the mighty Brabazon for the first time. There were few real problems despite its size, but its economics were poor and it was scrapped.

Much more promising was the Britannia, which he took to the air for the first time in 1952. In 1954 a major engine fire on the second prototype forced him to land on the mud-flats of the Severn Estuary.

Pegg retired from test flying in 1956 and held management posts until 1961.

PÉGOUD, ADOLPHE (1889-1915) – aerobatic pioneer and first to fly inverted. Within a few weeks of starting to fly, Pégoud was selected by Blériot as a test pilot for his company.

On 19th August 1913 he climbed out of an obsolete Blériot XI and descended by a parachute made by Monsieur Bonnet. The manoeuvres of the abandoned machine, which landed itself almost intact, gave him the idea of trying more ambitious piloted stunts.

He had a Blériot XI strengthened, and after having himself turned upside-down in the aeroplane on the ground to give himself the feel of the position, he flew inverted on 1st (or possibly 2nd) September 1913. He did so via the first half of a loop, and resumed normal flight by similar means. The aircraft was probably incapable of a full roll.

On 21st September he completed a loop, then believed to be another 'first'. Later it was found that the Russian **Nesteroff** had beaten him by a few days.

Pégoud became a French national hero and commanded huge fees for displays, although some criticised him as foolhardy. He was killed in action in 1915.

PEMBERTON-BILLING, NOEL (1881-1948) – founder of the Supermarine company. The eccentric Pemberton-Billing is said to have learned to fly within 24 hours for a bet. In 1913 he formed his aircraft company as an extension to his yacht-broking business. It was to be renamed Supermarine three years later. With his background, it was natural that he concentrated on marine aviation. His first flying-boat was the small PB1, but there is no record of it having flown. Further prototypes did fly, but the company survived mainly on licence and sub-contract work at first.

In 1916 he entered politics. To further his new career he pledged to build a 'Zeppelin-destroyer' so that civilians would fear those aerial raiders no more. The result was a remarkable machine, the P31E Nighthawk with two engines, four sets of wings, a then rare glazed cockpit, an upper gun turret and a gimbal-mounted searchlight in the nose. It was claimed it could patrol for 18 hours. It flew successfully but never went into production and left the Zeppelins unscathed.

Pemberton-Billing's most important role in history was establishing the company which went on to make the Spitfire and other fine aircraft.

PÉNAUD, ADOLPHE (1850-80) – flew models during the 1870s. Pénaud flew models of both helicopters and aeroplanes powered by rubber bands, later a method to be widely used on simple commercial models. They flew well, but more importantly, they were sold widely and helped to inspire many of the pioneers to come, including the Wrights. He patented a full-size monoplane with advanced features in 1876, but he attracted little but ridicule and took his own life.

PENROSE, HARALD (1904-) – Chief Test Pilot for Westlands. Penrose joined Westland in 1925 and spent his entire career with the company. He was Chief Test Pilot from 1931 till 1953.

His career was packed with incident. While flying a high-wing monoplane, the PV.7, the aircraft broke up. He escaped, with difficulty, but even after his parachute had opened his troubles were not over. He descended towards a railway, then "just as in a film", as he described it, a train appeared heading towards his landing point. Fortunately he landed just before the line.

The tail-less Pterodactyls designed by **Geoffrey Hill** gave more than their share of alarms. When Penrose was taxiing the Mark V he was dismayed when an entire wing collapsed.

The post-war Wyvern naval aircraft killed six test pilots and Penrose was fortunate, or skilled enough, not to have added to that number. One aircraft went into a roll at low altitude over Yeovil due to partial aileron failure. He saw all too clearly which house he was about to hit, but long experience and fast reactions enabled him to recover. Another Wyvern emergency led to a forced landing.

He continued flying when the company switched to helicopters, and became Sales Manager on his retirement from test flying. He has written a number of books, including the classic account of a test pilot's life, Adventure with Fate.

Harald Penrose seems to have attracted drama, even off duty. When sailing during the war he saw a fighter dive towards him and realised to his horror that it was a German. Just as he was braced for the stream of fire to end it all, the fighter turned away. He believed the pilot 'buzzed' him out of mischief.

PERCIVAL, EDGAR (1897-1984) – light aircraft builder. Australian-born Percival served in the wartime RFC, then the RAF, before turning to design, initially with the Saro company.

He formed his own business in 1932. His Gull was the first production British cantilever monoplane. Fast and strong, the various members of the Gull family set many pre-war records and fared well in racing. A war-time version, the Proctor, was built in large numbers as a trainer and communications aircraft.

Percival sold his company to the Hunting shipping group in 1944. He was not quite finished with aircraft design, for he had another dabble in the business with an aircraft built for agricultural work, the EP.9, in the 1950s. 21 were built, some under another company management, for Percival 'retired' again in 1958. He still conceived new designs but no more turned into hardware.

PETTER, EDWARD ('Teddy') (1908-68) - aircraft designer. A member of the Petter family who had, among other engineering interests, founded Westland, he designed the Lysander army co-operation aircraft, famed for its clandestine agent-dropping work, and the Whirlwind fighter.

He was a talented but sometimes eccentric designer. An instance was his insistence on running the Whirlwind engine exhausts through the fuel tanks, giving test pilot **Harald Penrose** an inevitable near-disaster. His Whirlwind went into service but its impressive appearance was not fully matched by its performance.

His greatest success came after he moved to English Electric, in the shape of his Canberra twin-jet bomber. The Canberra served the RAF for over 40 years and was widely sold throughout the world, including the United States where it was built by Martin as the B-57.

Petter handled early design studies for the Mach 2 fighter which became the Lightning. He left to join Folland in 1950, well before the Lightning flew (as the P1) in 1954. Design of that aircraft continued under Frederick Page.

He became Chairman of Folland following retirement of its founder. He tried to break the trend towards ever-more costly and heavier fighters by designing the little Gnat. He was disappointed the RAF never bought it as a fighter, but it made an excellent advanced trainer and as such delighted airshow crowds as the mount of the Red Arrows for many years. It did still prove itself as a fighter, for it was built in India in large numbers and acquitted itself well in action.

Petter spent his last few years as a recluse in Switzerland.

PIERSON, REX (1891-1948) – designer for Vickers. An engineering apprentice with the company, Pierson became Chief Designer in 1917, holding the post for 28 years. He did much of the design work on the Vimy, famed for its transatlantic flight and two other trailblazing epics.

In World War Two his greatest contribution was the Wellington, designed in conjunction with **Barnes Wallis**, whose sturdy geodetic structure was used. Wellingtons were well-liked by crews because of the amount of damage they could take. 11,460 were made, the highest of any British bomber. Pierson used it as the basis for his post-war Viking airliner, although in the end it became virtually a new airframe. 163 were made, a modest success, and in addition there were military transport and trainer versions. Pierson retired in 1945.

PHILLIPS, HORATIO (1845-1926) – pioneer of aerofoil theory and practice. Phillips took out patents for aerofoil sections as early as 1884. He studied the theory of lift scientifically, even building a form of wind-tunnel. From his research he understood the need for curvature on both upper and lower wing surfaces, putting him far ahead of most contemporaries.

He is believed to have achieved gliding flights in the 1890s, and a steam-powered model flew in 1893. Early in the 1900s he turned to adding engines and propellers of his own design. In 1907 he is reputed to have become airborne for 500 ft (150 m) in one of his 'multiplanes', an aeroplane with 20 narrow wings mounted like a Venetian blind. It has been claimed as the first powered flight in Britain, but is not widely recognised as such as the machine lacked real control and there was insufficient evidence for his achievement.

PIASECKI, FRANK (1919-) – twin-rotor helicopter specialist. A small company was formed in 1940 to promote helicopter research. The first Piasecki was a small single-rotor machine flown in April 1943.

His long line of twin-rotor helicopters started with the HRP, first flown in 1945 and generally known as the 'Flying Banana' due to its cranked fuselage. It saw long service with the American services, including valuable rescue work.

Piasecki had to relinquish his control of the company in the 1950s and in 1960 it became the Vertol Division of Boeing. Piasecki's twin tandem rotor layout proved a sound one and was perpetuated in the Chinook, first flown in 1961 and still one of the most widely-used machines in its class.

PILCHER, PERCY (1867-99) – Scots gliding pioneer. Pilcher had long been interested in aeronautics and is believed to have been further inspired by **Lilienthal's** early glides in 1891. He built his first glider, the *Bat*, in 1895, but before flying it he visited Lilienthal for advice and to fly his gliders. On his return, he flew the *Bat* a number of times in Dumbartonshire with fair results.

Further gliders followed, the *Beetle*, a rebuilt *Bat*, and the *Gull,* without achieving much advance on the original.

He moved to Kent and in 1896 built his last and best glider, the *Hawk*. On one flight he covered 750 ft (230 m). An advance on other pioneers of the time was his use of a wheeled undercarriage. He could be launched using a long fishing-line, like a modern glider winch-launch apart from the more literal horse-power used by Pilcher. On 30th September 1899 a bamboo tail member of the *Hawk* failed in flight and Pilcher died later.

He had been building a 4 hp engine with a partner, Walter Wilson, and he could have been nearing the stage of progressing to powered flight. However, the planned engine power seems low for success, and he lacked the control developed by the Wrights. He is, nevertheless, an under-rated pioneer.

PIPER, WILLIAM (1891-1970) – founder of the Piper Aircraft Corporation. Piper took over a small light aircraft manufacturer which had run into financial trouble in 1930. He went on to make the company one of the world's leading builders of light aircraft, reaching a total of 100,000 by 1977.

The Piper Cub, which first flew in 1930, became a classic. Most American wartime generals travelled around their units in Cubs, and it continued to be made in large numbers after the war.

PORTE, JOHN (1884-1919) – flying-boat builder. Porte transferred his interests from travelling below the sea to flying above it, for he had been a submariner before turning to flying. He built a two-seat glider in 1909 and flew it from Portsdown Hill, with the usual share of mishaps.

He progressed to making a Santos-Dumont style monoplane, and for a while represented the Deperdussin company in Britain. He competed at flying events, on one occasion creating a 'sensation' by exceeding 90 mph (145 kph).

In 1913 he worked with **Glenn Curtiss,** starting Porte's interest in flying-boats. Ambitiously, he planned a transatlantic flight, via the Azores, before war intervened.

Porte persuaded the Admiralty of the value of flying-boats and secured orders, first from Curtiss then for versions of his own. After the war he would be accused unjustly of improper payments for the Curtiss purchases. In 1916 he built the then largest flying-boat in the world, which he named, tongue-in-cheek, the *Baby*. One was used for a 'piggy-back' trial of launching a Bristol Scout in flight, preceding by 20 years the better known *Mercury-Maia* scheme.

Porte not only built flying-boats, he flew them on wartime patrols too. This is especially remarkable since he had suffered from tuberculosis since before the war, and finally succumbed to the disease in 1919.

POST, WILEY (1898-1935) – flew twice round the world, with one eye! As a spectator at a barnstorming event in 1919, Post volunteered to take the place of a parachutist who had been injured. He stayed on the barnstorming circuit for a while, but continued his regular job on the oilfields as well. There he sustained an injury which lost him use of an eye.

In 1930 he became personal pilot to an oil executive, flying a Lockheed Vega *Winnie Mae*. In the same year he won a race in this aeroplane from Los Angeles to Chicago.In June 1931 he flew *Winnie Mae* round the world with navigator Harold Gatty, setting a record of 8 days, 16 hours. This was despite being bogged in soft ground at one point and later damaging the propeller.

Two years later he repeated the flight, this time solo. By now the autopilot had been developed and he had one fitted in *Winnie Mae*. This time it took him just 7 days, 19 hours.

Post became interested in high-altitude flight and equipped with pressure suit, engine supercharging, and a jettisonable undercarriage he flew at up to 55,000 ft (16,800 m). Landing was on a skid, with the propeller stopped horizontally. Presumably the owner of *Winnie Mae* was tolerant about use of his aeroplane.

In 1935 he set out with actor and humourist Will Rogers on a long-distance flight in a Lockheed Orion. In Alaska they spun into the water soon after take-off, it was believed due to engine icing, and both were killed. His faithful *Winnie Mae* is in the Smithsonian, Washington.

QUILL, JEFFREY (1913-96) - test pilot. Rated as 'exceptional' early in his RAF career, in 1934 Quill faced a challenge when leading the Meteorological Flight - could his team fly every scheduled sortie over a 12 month period? It had never been done before. Doggedly they flew through every type of weather, 'bending' a few of their Siskins, but not one mission was missed.

He joined Vickers as a test pilot in 1935. By then the company had bought Supermarine and he first flew a Spitfire in March 1936 and for the next 12 years handled the major share of its development flying. He flew other Vickers types too, and when spinning a Wellesley was forced to bale out when it failed to respond. He saw it land on a house, fortunately without harm to the occupants. On another occasion he had just taken off in a Vickers Venom fighter when the engine failed with nowhere ahead to land. After an anxious few seconds the engine recovered.

One of his contributions to test flying was to set out a proper schedule of flight tests for production aircraft. Hitherto it had been haphazard with no list of checks to be done.

To ensure he understood what squadron pilots really needed, he joined the RAF in 1940 to fly in combat. Later he did some naval flying to help overcome handling problems on the Seafire, the carrier version of the Spitfire. He found the experiences invaluable in feeding priority actions to the manufacturer.

Quill retired from test flying in 1947 and went on to hold sales appointments in Vickers and its successors, including Panavia, the builders of the Tornado. He retired in 1978. He was a founder of the Spitfire Society and wrote a number of books on the aircraft.

> **Just Dropping in**
>
> ***Years after his Wellesley incident, Quill's daughter was approached by a girl who asked if her father was a test pilot. When his daughter confirmed it, the girl added "Well, he dropped a damned great aeroplane on my mother's house!"***

READ, ALBERT (1837-1967) – first to fly the Atlantic. Read was a naval officer who started flying in 1915 after eight years at sea. On 16th May 1919 he commanded Curtiss flying-boat NC-4 when it left Newfoundland for Europe via the Azores. The NC-4 was one of three flying-boats making the attempt - a fourth had withdrawn.

The US Navy had strung out 41 destroyers along the route in case of mishap, suggesting a low level of confidence. The precaution proved wise, for two aircraft came down in the sea. Read's aircraft alone reached Lisbon, and continued on to Plymouth, a story of skill and endurance even if eclipsed by the non-stop Vimy crossing a few days later.

> **More Boat than Flying-Boat**
>
> ***The crews of the two flying-boats which alighted en route were rescued, but one kept afloat through 60 hours of gales until it taxied into the Azores, an epic in itself.***

REITSCH, HANNA (1887-1956) – legendary woman test pilot. Hanna Reitsch was an accomplished pre-war glider pilot, winning a handful of world records. She started test flying in the late 1930s.

She came to world-wide notice when she set a helicopter distance record of 67.7 miles (109 km) in October 1937 in a Fw61. Even more publicly, she demonstrated it in 1938 inside the Deutschlandhalle in Berlin.

She tested a vast range of wartime types, and undertook hazardous trials of cable-cutting devices. She flew the great Messerschmitt 321 Gigant glider, not surprisingly finding its controls heavy. Before one flight, on which she was not earmarked as the pilot in command, the captain prevented her boarding. Perhaps he had a sense of foreboding, for minutes later the glider and its three tugs crashed with the loss of 129 lives.

She was badly injured testing an unpowered prototype of the Me163 rocket fighter, flying it as a glider. The Me163 was undamaged but her injuries were caused by not strapping in properly and because she had deployed the gunsight in front of her face, although no guns were fitted.

She recovered and tested a manned version of the V1 flying bomb. As well as collecting data for the pilotless missile, there was a proposal for a suicide manned version.

In the last days before Berlin fell she flew a Fieseler Storch into the city under Russian fire, carrying the general who was to succeed **Goering**. She was one of the last to see Hitler.

She resumed gliding after the war and set a new German record in 1970.

RENTSCHLER, FREDERICK (1887-1956) – founder of the Pratt and Whitney engine company. In 1925 Rentschler, then with Wright Aeronautical, felt that the owners of the company were allowing development to stagnate. He left to form an aero-engine division of Pratt and Whitney, machine tool manufacturers.

The new company concentrated on air-cooled radials, first with the Wasp and then under designer 'Luke' Hobbs the R-1830 Twin Wasp, one of the most successful aero-engines ever made, and still flying in some numbers in Douglas DC-3s. He also took Pratt and Whitney into the jet age with his JT3 series.

RICHTHOFEN, MANFRED VON (1892-1918) – legendary German fighter pilot. Von Richthofen started as a cavalryman but applied for a flying post when trench warfare proved too static for his liking. In 1915 he was posted as an observer, then often senior to the pilot.

In 1916 he joined one of the new fighter units and soon his skill gave him a fearsome reputation, later enhanced by his practice of flying distinctive all-red aeroplanes. He was popularly called the 'Red Baron', although he held no such title in nobility.

Von Richthofen scored 80 victories. His courage and skill are beyond question, but unlike most fellow pilots he had a sadistic streak and revelled in the fate of his victims. Nor was modesty one of his qualities – he bought himself a silver cup after each victory. He was killed on 21st April 1918, his demise usually being credited to a Canadian, Roy Brown.

His brother Lother was no mean pilot either, logging 20 victories. He was killed in an accident in 1922.

RICKENBACKER, EDWARD ('Eddie') (1890-1973) – top American fighter pilot of World War One, and head of Eastern Airlines. Before starting flying, Rickenbacker (originally Reichenbacher – he changed it to avoid anti-German feeling) was a racing driver, and in 1916 held the world speed record at 134 mph (216 kph).

Rickenbacker, Edward *(continued)*

When America entered the war he applied for flying training, and after several refusals received an operational posting early in 1918. He won 26 victories, the highest for any American.

After the war he built Rickenbacker cars until the firm failed in 1927, and ran the Indianapolis motor race. Back in aviation, he joined Eastern Airlines and was its president until his retirement in 1963. Like many strong-minded men, he both wielded respect and made foes.

During the Second World War he handled some military work and once ditched in the Pacific, spending 22 days on a raft. His companions attributed their survival to his leadership.

RODGERS, CALBRAITH (1879-1912) – first to fly across the United States. In response to a $50,000 prize offered by the newspaper magnate Randolf Hearst for a coast-to-coast flight in under 30 days, Rodgers set off from New York on 17th September 1911. His Wright biplane was named **Vin Fiz** after a drink made by his sponsor.

He reached the Pacific after an extraordinary epic of crashes and rebuilds. Rodgers may not have been the most skilled of pilots but he never lacked determination, for the flight took 84 days. That's 30 days more than the record for walking the distance! It was said the only original parts to reach the end were the rudder and oil pan.

In the following year he was killed when his aircraft hit a gull.

ROE, SIR ALLIOT VERDON (1877-1958) – first 'all-British' flight, founder of Avro and co-founder of Saunders-Roe. During an adventurous career in Canada and South Africa, Roe became interested in flight and studied birds, notably the albatross. He was also yet another aviation pioneer who had been a racing cyclist, and his early flying trials were largely funded by his prize winnings.

In 1907 he won a model aircraft competition run by the *Daily Mail,* and in 1908 advanced to a full-scale biplane. In it he made several 'hops', and on 8th June 1908 he may well have made the first flight by a British national, but there were no official observers present to verify it.

Success came on 13th July 1909, when he achieved a proper flight in his triplane *The Bulls-Eye,* or more irreverently known as the *Tripehound.* This was the first flight by a British national in a British aeroplane. In 1910 the triplane was destroyed en route by train to a flying meet at Blackpool – it was burnt out due to sparks from the locomotive. He managed to assemble another in three days and still took part in the flying! Aircraft building takes rather longer nowadays.

He built the first cabin aeroplane and was one of the first to use the now-traditional stick and rudder controls. His most important creation was his Avro 504, over 5,000 of which were made, and possibly many more including large numbers in Russia.

In the late 1920s Avro were in financial trouble and Roe lost control of the company, which recovered to become one of the great names in aircraft building. He teamed with **Sam Saunders** to form Saunders-Roe, or Saro. The new company made a line of fine flying-boats, and, it must be said, some poor ones including the hapless wartime Lerwick.

Roe, who changed his name to Verdon-Roe, spent his last few years building what he called a 'two-wheeled car', although how it differed from a motor-scooter was not obvious.

Avro's Classic Trainer

Roe flew his little 504 biplane before the First World War. In action it was used as a 'scout', and even as a bomber. A raid by three Avro 504s on the Zeppelin works on 21st November 1914 is considered the first ever strategic bombing raid.

However it is as a trainer that the 504 was best known and it was built and used as such for over 20 years. A few were still flying into the Second World War, but not, mercifully, in the front line!

ROHRBACH, DR. ADOLPH (1889-1939) – pioneer of all-metal cantilever wings. A designer of wartime bombers with Zeppelin-Staaken, in 1920 Rohrbach built his E.4250 four-engine airliner of visionary concept. With its cantilever wing and all-metal construction it would have looked modern 15 years later.

Progress was halted due to restrictions on aviation activity in Germany, but he moved to America and his work inspired **Northrop** and **Boeing** in their ventures into advanced structures.

When Power Gave Way to Sail

A flying-boat built by Rohrbach carried unusual equipment for an aircraft – a mast and sail! If the machine was stuck on the water without power, out would come the mast and off they would sail, or so they hoped.

ROLLS, THE HON. CHARLES (1877-1910) – co-founder of Rolls-Royce. Rolls was a pioneer motorist, a racing driver, balloonist, and (yes, another!) racing cyclist. In 1904 he was selling foreign cars but wanted to add a competitive British vehicle to his range. He was introduced to **Henry Royce** in a historic meeting in a Manchester hotel and drove the prototype car. He was so excited by its silence and smoothness that when he took it to London he woke his business partner, Claude Johnson, at midnight to test the car. He undertook to take all Royce's output, to be sold under the name Rolls-Royce.

The Rolls-Royce company was founded in 1906, moving to Derby in 1908, the location being settled by an offer of low electricity prices.

In 1908 Rolls met **Wilbur Wright** in France and was smitten by flying, so much so that his interest in cars and the company waned. He ordered a Wright biplane and on 2nd June 1910 made the first two-way crossing of the English Channel.

On 12th July 1910 he was taking part in a spot-landing competition at Bournemouth. As he pulled up from his descent part of the tail assembly failed, probably due to a modified part he had fitted the day before. His was the first aeroplane fatality in Britain.

ROYCE, SIR HENRY (1863-1933) – co-founder of Rolls-Royce. It is wholly appropriate that this entry and the above are consecutive. Many have remarked how well the two names blend together, particularly in connection with the luxury cars.

Royce started his first engineering business as FH Royce in 1884, making electric bells, dynamos, and later, cranes. This business continued until 1933. Some of these quality products are still in use.

He built his first car in 1904. He may have done so because the electrical business was being hit by cheaper, but less durable imports, or it may have been dissatisfaction with cars he saw. The meeting with Rolls and starting the company are related above.

By 1910 years of overwork and neglect of his diet had caused his health to collapse and he nearly died. On medical advice he set up homes in Sussex and the South of France, with design offices at each. He left Derby never to return, although the factory is still 'Royce's' in the city.

On the outbreak of war in 1914, Royce was asked to build aero-engines. His first design, the Eagle, was a water-cooled V-12, a formula perpetuated in the famous Merlin. By a prodigious effort it was ready within six months of starting work.

Most engines of the day fell short of their forecast power (a problem by no means unknown today!), but on its first run the Eagle roared past its nominal 200 hp to reach 225 hp.

Two other wartime engines followed, the Falcon, essentially a scaled-down Eagle, and the little Hawk, used in small airships. All the engines were popular with aircrew, for they rarely let them down, literally or otherwise, even if they did sometimes leak water and need repair with chewing-gum. It has been claimed that 60% of British aircraft at the end of the war were Rolls-Royce powered. After the war Eagles powered the Vimys which blazed trails across the Atlantic, to Australia, and most of the first flight to South Africa. Many early airliners also used Eagles.

In 1925 came the Kestrel with cylinders mounted in a single block, and in 1929 and again in 1931 Royce and his team wrung spectacular powers from the special racing R engines to win the Schneider Trophy races and to set world air speed records. For good measure the R engines broke the land and water speed records as well!

In 1932, when Royce was nearing the end of his life, he authorised start of work on a new engine at company expense, the PV12. He did not live to see or hear the new engine run, but this last decision was a momentous one for the world - the PV12 was to become the most important aero-engine of all time, the Merlin.

The durability of Henry Royce's creations is legendary and just one example will suffice. When a replica Vimy was built in 1960 for a film of the Atlantic flight, a search for Eagle engines turned up two, still working away pumping water in Holland. When tested, their power was within 2 hp of their design output. Sadly the replica was destroyed in a ground fire.

Despite starting the line of many of the finest aero-engines, it is believed Henry Royce never made a single flight.

Pilâtre de Rozier – first man to fly

ROZIER, FRANÇOIS PILÂTRE DE (1757-85) – first man ever to fly. A scientist who had the invention of a breathing apparatus to his credit, de Rozier became interested in balloons when he saw the **Montgolfiers'** livestock-carrying ascent. When a manned balloon was mooted, King Louis XVI reputedly decreed that it be flown by criminals because of the risk.

De Rozier *(continued)*

De Rozier was outraged that the honours for such a historic event might go to those so unworthy and via the **Marquis D'Arlandes** persuaded the king to allow him to fly the balloon.

On 15th October 1783 he started tethered trials, gradually improving his proficiency in this new art over the next few days.

The first free flight by mankind took place on 21st November 1783 from the Chateau la Muette in the Bois de Boulogne, near Paris. Not surprisingly, 'a vast multitude' were present when de Rozier and the Marquis d'Arlandes rose into the air in the highly decorated balloon. They doffed their hats as they rose, but the flight was far from serene, for the fire needed stoking, and more seriously, kept burning holes in the fabric.

Disaster was ever close, but averted with sponges and buckets of water. They landed after 25 minutes, confounding all those who had predicted untold perils in the sky. The age of flight had begun.

De Rozier made further ascents, including a 36 mile(57 km) flight, and a hazardous one in a colossal 'Montgolfière' of 700,000 cu ft (19,500 cu m). Not least of the perils was a dispute with swords drawn before launching about to how many could accompany him.

In 1785 he attempted an English Channel crossing. The balloon was a hybrid affair comprising a hydrogen envelope above a hot-air balloon. It sounded an unwise combination, and so it proved.

According to some reports he had doubts about it, but duly cast off on 15th July 1785 with its builder, Jules Romain. The almost inevitable happened and the balloon plummeted on fire from around 5,000 ft (1,500 m).

RUSSELL, SIR ARCHIBALD (1904-95) – Chief Designer, Bristols, and a prime mover of Concorde. Russell joined Bristols in 1926, succeeding **Leslie Frise** as Chief Designer in 1944. He took over design of the Bristol 170 Freighter, a fine workhorse, the Brabazon, and the Britannia. This elegant airliner was plagued by delays in development and never reached its sales potential.

He was one of the architects of Concorde, crucially arguing that supersonic travel made sense only on long routes, whereas the French wanted to build a medium-range transport. The design was a technical triumph and does all that was ever claimed for it, other than make a profit on a normal accounting basis.

He became Chairman of the Filton division of the British Aircraft Corporation in 1968 and retired in 1971.

Discordant Concorde

Concorde was always controversial – there was even argument about whether or not to spell it with the final 'e'. Nevertheless the enormous technical task of carrying 100 passengers at twice the speed of sound was accomplished successfully. The task took longer than foreseen and costs rose.

Meanwhile the original yardstick of matching Boeing 707 economics became out of date when the Boeing 747 started service in 1970 with far lower costs per seat, and rapidly rising fuel prices hit Concorde harder than the subsonics.

As a result only 14 Concordes entered service.

RUTAN, BURT (1943-) and **DICK** (1938-) – designer of radical homebuilt aircraft, and the first to fly around the world without refuelling. Burt Rutan has designed a series of unconventional but most efficient light aircraft which are sold as plans for purchasers to build at home. Most are of canard, or tail-first layout, and the materials are as radical as the appearance. Several thousands of plans have been sold.

In 1986 Burt Rutan, brother Dick, and Jeana Yeager built the Voyager, an aeroplane designed to fly around the world without refuelling. It had two engines, one in the nose for the early part of the flight only, and a second in the tail to run continuously. The tips of its 111 ft (33.8 m) span wing could flex through a range of 60 ft (18 m), a degree of bending which looked so frightening that the crew resolved not to look at the wing-tips in flight!

On 14th December 1986 Dick Rutan and Jeana Yeager set off on their epic flight, returning after nine days, a remarkable trial of endurance, and of crew compatibility! Their greatest problems were fatigue and avoiding storms, for severe turbulence would probably have been fatal to Voyager. Despite modern technology, the risks were comparable to those faced by the trailblazers of the 1930s. As one of the team said "it was the last first in aviation."

As the rear engine was intended to run without stopping, to save weight no starter was fitted. The one start was by hand prop-swinging, or as Dick Rutan called it 'The Hemingway starter' - one slip and it's "farewell to arms".

SADLER, JAMES (1751-1928) - first British national to fly. The first ascent in a long ballooning career took place on 4th October 1784 at Oxford. He used a 'MontgolfiËre' hot-air balloon, but all his later flights employed hydrogen.

Some of his flights were far from uneventful and included at least three heavy landings followed by loss of the balloons, and two ditchings at sea. For a time he forsook flying to design steam engines, but the lure of the air was too great and he was airborne again in 1810. Once more life was eventful, and on one flight he averaged 80 mph (120 kph) over the ground.

Few would have flown in such a wind, even if it was less at ground level, and predictably the landing was exciting. Sadler was thrown out leaving his terrified passenger to rise again, but all ended well.

He almost completed an Irish Sea crossing in 1812 but once more ended in the sea. His son, Windham, did manage this feat in 1817, but was killed ballooning in 1824.

SANTOS-DUMONT, ALBERTO (1873-1932) – airship pioneer and first to fly an aeroplane in Europe. Soon after arriving in Paris from his native Brazil, Santos-Dumont was introduced to ballooning. In 1898 he built the first of his 14 airships.

Santos-Dumont– Brazilian pioneer of European aviation

He was fortunate to have ample funds from family coffee estates. His first flight in 'No.1' ended in trees when he yielded to advice and took off downwind. He learned the lesson and on 20th September 1898 he made a proper flight, although loss of gas led to an undignified landing.

On 19th October 1901 he flew No.6 from outside Paris, circled the Eiffel Tower and regained his starting point. It won him a 100,000 franc prize, which he gave away. He became a popular figure in Paris, using his No. 9 as a 'runabout', often tethering it outside houses or restaurants.

On learning about the Wrights' achievements he built his No.14 bis aeroplane, continuing the airship numbering sequence. This tail-first design was flown standing up. To modern eyes it looks as though he was flying it backwards. On 23rd October 1906 he made the first aeroplane flight in Europe in this improbable machine. Some authorities class this as a 'hop' and regard his first sustained flight as taking place on 12th November.

In 1908 he built his first Demoiselle monoplane, in which the pilot sat below the wing on a platform above the wheels. Using a bamboo structure, it was powered by a two-cylinder motorcycle engine. Although demanding to fly, it was successful and a number were made, but with stronger materials than the bamboo.

In 1910 his health deteriorated and he returned to Brazil. He kept in touch with aviation, but sadly this apparently most carefree of men became depressed at the military use of flying and took his own life.

SAUNDERS, SAMUEL (1856-1933) – co-founder of Saunders-Roe. A member of a long-established family of boat-builders, Saunders patented a new method of hull construction which was widely used by marine designers from 1909 onwards.

Saunders built aircraft under licence during World War One, branching out into his own designs in 1917. Few progressed into production until he joined forced with A.V. Roe in 1928 to form Saunders-Roe, or Saro.

Under the new joint management two successful flying-boats were built in the 1930s, the London and the amphibious Cloud. The company continued into post-war years and in 1959 built the world's first hovercraft.

A Saro Cloud amphibian being used as a navigation trainer, with windows covered so the trainees were unable to cheat by looking outside, landed on rough water. One pupil believed they had landed on a ploughed field and that evacuation was in order. He promptly leaped through the door – straight into the sea.

SAYER, GERALD ('Gerry') (?-1941) – flew first British jet. After a background as an RAF test pilot and a short spell at Hawker, Sayer became Chief Test Pilot at Glosters in 1934.

On 15th May 1941 he made the historic first flight of the Gloster E28/39 jet research aeroplane from Cranwell. Due to poor weather the flight was delayed until 7.35 pm – the tension for all concerned must have been hard to bear. Not until after the war was it realised that the Heinkel 178 had flown in 1939.

Sayer was killed later that year when a Typhoon he was flying collided with another in cloud.

SCHNEIDER, JACQUES (1879-1928) – started the Schneider Trophy races. A lover of anything fast with a spice of danger, Schneider was an early pilot, balloonist, and power-boat racer. It was his interest in the sea which prompted him in 1912 to donate his famous trophy for seaplanes.

It proved a spur to aviation far beyond anything he could have foreseen. It was last contested in 1931, for under his rules any country winning it three times in succession would keep it, and that year was Britain's third victory in a row.

Schneider did not live to see the final contest or the benefits they would bring, for he died suddenly in 1928.

SCOTT, CHARLES (1903-46) – record-breaker and racing pilot. Scott joined the RAF in 1922, moving to Australia in 1927 where he became an airline pilot with Qantas.

He took his first record in 1931 by flying a Gypsy Moth from Britain to Australia in 9 days. He was to take the record for this route twice more. His return flight a month later notched up his second record.

His greatest triumph was winning the 1934 MacRobertson race to Australia, flying a de Havilland Comet racer with **Tom Campbell Black**. They flew through a fearsome tropical storm over the Bay of Bengal, needing, as they put it, "all four feet on the rudder pedals." Over the Timor Sea there was further anxiety when an engine had to be shut down due to oil pressure problems. They reached Melbourne triumphantly, but at the limits of exhaustion. Scott called it the worst trip he ever experienced. The fact that the two pilots were of similar experience added to the stresses.

In 1936 he joined with a joy-riding pilot, Percival Phillips, in buying **Alan Cobham's** 'Flying Circus', but the summer that year was a poor one and the venture ended in the autumn.

More successful was his victory that year with co-pilot Giles Guthrie (30 years later a controversial chairman of BOAC) in the Schlesinger Race to South Africa. Their Percival Gull was the only one of the nine competitors to complete the race. Scott's experience in taking enough sleep may have been a key to success where others failed, and in some cases died.

As sometimes happens to men of action after a life of excitement, he found it hard to settle into quieter ways and took his own life in 1946.

At the end of his first record flight to Australia, Scott was invited to open a hospital. The car taking him there collided with a cyclist who was slightly hurt and taken to another hospital. In his speech, Scott explained his lateness and hoped he would not fill the new unit with his victims. Silence greeted the light-hearted remark. Later he realised why – it was a maternity unit.

SCOTT, HERBERT (1888-1930) – airship commander and first to fly the Atlantic both ways. Scott started his airship career flying naval non-rigids in the First World War, and he also commanded the first British rigid, HM Airship No.9.

On 2nd July 1919 he left East Fortune, Scotland, in airship R-34, reaching Mineola, Long Island, on the 6th. The return flight, to Pulham in Norfolk, took place between 10th and 13th July. He thus became the first to fly the Atlantic westbound and the first to make the double crossing. To many this flight signalled that the airship would dominate long-haul air travel.

Scott devised the high mast for mooring large airships, avoiding the need for massive handling parties on the ground.

Rightly acclaimed for his 1919 flights, a series of mishaps later gave him a reputation for poor judgement. Despite this, his experience won him a place on the inaugural overseas flights of both the R-100 and R-101, but his seniority over the actual captains caused strains in command. He died in the R-101 disaster.

The famous naturalist, painter and sailor Sir Peter Scott somehow managed to add gliding to his already full life, and remarkably in view of all his interests, he was National Champion in 1963.

SCOTT, SHEILA (1927-87) – set over 100 light aircraft records. Sheila Scott learned to fly in 1959 and lost little time in entering competitions, for she flew in the King's Cup air race the next year.

Between 1964 and 1972 she set over 100 records, in most cases in single-engine Piper Comanche aircraft. On 18th May 1966 she set forth on a round-the-world flight, experiencing a worrying spell over the Pacific when fuel leaked into the cabin. She reached home on 20th June.

She switched to a twin-engine Piper Aztec for a flight over the North Pole in 1971. For record purposes the flight was from 'equator to equator.'

Her list of records is a tribute to careful preparation and determination, not just in the air but also in the grind of seeking sponsorship and funds. It was not flying which killed her, but cancer.

SEGUIN, LOUIS (1869-1918) and **LAURENT** (1883-1944) – founders of the Gnôme engine company. The French brothers started making car engines from 1905 and extended their range to aero-engines two years later. They chose the rotary type, in which the entire engine rotates around the crankshaft to increase cooling. It worked surprisingly well and a large proportion of First World War engines were of this type, with Gnômes well to the fore. In post-war years the static radial displaced the rotary, being more suitable for larger engines.

Thomas Selfridge earned an unwanted place in history, as the first aeroplane fatality. On 17th September 1908 Orville Wright was demonstrating a biplane to him during trials for the US Army when a propeller blade failed, damaging the structure. Orville was injured.

SEVERSKY, ALEXANDER (1894-1974) – originator of the Republic Aviation Corporation. A Russian combat pilot of World War One, Seversky lost a leg but returned to action. He was in America at the time of the Revolution and elected to remain there. In 1931 he formed Seversky Aircraft Corporation. Fellow Russian Alexander Kartveli became the designer.

The company was renamed Republic in 1939 and produced one of the principal American wartime fighters, the massive P-47 Thunderbolt, widely known as the 'Jug'. Over 15,000 were made, some serving with the RAF.

Richard Seys performed a feat in 1943 which has never been repeated, of flying a glider across the Atlantic on tow. With co-pilot Fowler Gobell he flew a Waco CG4A behind a Dakota in stages of up to seven hours.

Handling a heavy cargo glider for such periods, much of the time in cloud, was a real test of fortitude. It was clearly not a suitable idea for routine runs.

SHARMAN, HELEN (1963-) – first Sharman, Helen (1963-) – first British national in space. Helen Sharman was one of 13,000 applicants who responded to an advertisement to fly on a Soviet spaceflight, the Juno mission.

Launch was on 18th May 1991 in the Soyuz TM12 spacecraft. She and her fellow crew members visited the Mir space station where they spent a week performing scientific experiments.

SHEPARD, ALAN (1923-) – first American in space. The launch of America's first manned spaceflight took place on 5th May 1961. The Mercury capsule, *Freedom 7*, was launched by a Redstone rocket. The flight lasted 15 minutes and reached a height of 116 miles (187 km). Unlike Yuri Gagarin's flight the previous month, this mission was not intended to enter orbit. However, in one sense it was an advance on the Russian spaceflight in that Shepard was able to practice controlling the Mercury capsule.

For medical reasons Shepard was unable to fly again in space until 1971, when he commanded *Apollo 14.* the third moon landing.

SHOLTO DOUGLAS, WILLIAM (Lord Douglas of Kirtleside) (1893-1969) – RAF leader and airline chairman. A successful pilot and squadron leader in the First World War, Sholto Douglas became Chief Pilot with Handley Page Air Transport in 1919. He soon returned to the RAF and held senior appointments in the service until 1947. He succeeded **Dowding** as head of Fighter Command in 1940, and later led Middle East Command and Coastal Command.

In 1948 he became a director of BOAC. In the same year he became Lord Douglas. In 1949 he was appointed Chairman of BEA, a post he held for 15 years. He saw the transition into the turbine era, becoming the first operator of the Viscount in 1953 (the prototype had flown passenger services for a month in 1950). In a gesture which gave a boost to the airliner's sales, he allowed the American airline Capital to take three of BEA's positions on the production line, giving the Viscount its vital breakthrough into the United States market.

SHORT BROTHERS, HORACE (1872-1917), **EUSTACE** (1875-1932), and **OSWALD** (1883-1969) – British aircraft builders. The beginning of the long involvement of the brothers in flying was a balloon ascent by Oswald in 1898. By 1900 he and Eustace were making balloons. Horace was living a life of adventure travelling in remote areas of the world. Due to childhood meningitis he had a distended head, but clearly it had not impaired his brain. He had a number of inventions to his credit before joining his younger brothers in 1908. In that year they embarked on aeroplane building, with licence-built Wright biplanes. They were handicapped by finding that the Wright brothers had no drawings of their biplane! Horace measured the machine himself.

By 1911 they had started on their own designs, including an early twin-engine type. The early aviator Sir Frank McClean bought some of the first Shorts, and helped the brothers financially. With these aircraft he started British naval aviation.

The naval association continued with a line of biplane seaplanes in World War One. A further activity was airship building, including the small 'blimps' (virtually an airship envelope with an aeroplane fuselage suspended below) and later some large rigids.

Horace died before the end of the war, at which time military orders dried up and the company turned to boat and vehicle work to survive. Boldly, Eustace and Oswald built an all-metal biplane, the Silver Streak, but regrettably official indifference killed this far-sighted project.

In 1932 Eustace ran into the shore after alighting in a Short Mussel seaplane - he had suffered a heart attack.

Under Oswald's leadership the company built the magnificent 'Empire' flying-boats, some of the most advanced aircraft in the world in the 1930s, their military derivative, the Sunderland, and the Stirling bomber. Oswald retired upon nationalisation of the firm in 1943, but remained Honorary President until his death.

SIKORSKI, IGOR (1889-1972) – pioneer of large aircraft and of helicopters. Sikorski started his long career in aviation by building a helicopter in 1909. It failed to fly, but he was to return to this form of flight later. He turned to fixed-wing aircraft in 1910, soon thinking on a large scale and building *Le Grand* in 1913. It was the world's first four-engine aircraft, and was also advanced in having an enclosed cabin. It was destroyed in a freak accident when a Voisin biplane broke up above it and the engine fell on *Le Grand.*

Even more ambitious was his *Ilia Mourumetz,* which set a world record in 1914 by carrying 16 people and a dog. 73 of the type were built, most serving as bombers.

Sikorski left Russia after the Revolution and he settled in America, forming his new company in 1923. Until 1940 it was best known for a series of fine flying-boats.

He tried again with helicopters in 1939 and on 13th May 1940 achieved free (non-tethered) flight in his VS-300 which used the now classic main and tail rotor arrangement. The problems of control proved formidable – at one time it would move in any direction but forwards – but perseverance was rewarded and Sikorski became a world leader in rotary-winged flight. He formally retired in 1957 but remained a consultant to the end of his life.

> ***Ilia Mourumetz included a rare feature on an airliner - an outside promenade deck above the rear fuselage. It must have taken some fortitude, or Vodka, to have used it in a Russian winter.***

SLINGSBY, FREDERICK (1894-1973) – sailplane builder. Fred Slingsby, a former wartime pilot, was flying gliders in the 1930s when **Philip Wills** asked him to build a sailplane. Slingsby Sailplanes was formed in 1934.

The golden age for Slingsby came in the 20 years after the war. Many world records and championships fell to his products, which were exported all over the world – in 1959 a Skylark was the first British aircraft sold to Soviet Russia. The pre-eminence was lost when exotic designs in new materials eclipsed the wooden Slingsby designs, and a fire at the works set the company back further. Ironically the firm now makes a highly successful light *powered* aircraft.

> ***Nowadays it would probably be unthinkable to name an aircraft after a cigarette, but in the 1960s the Slingsby T49 was named Capstan, needless to say in a sponsorship deal.***
> ***Perhaps more 'comical' was the earlier naming of the T42 two-seater. In association with a newspaper for young readers it was named*** **Eagle.**

SMITH, SIR KEITH (1890-1936) and **SIR ROSS** (1892-1922) – first to fly from Britain to Australia. Both brothers served in the air during World War One, in a fortunate combination Ross as a pilot and Keith as an observer and navigator.

In 1919 the Australian government offered a £10,000 prize for the first flight by Australians from Britain to their own country. The journey time had to be within 30 days.

On 12th November 1919 the Smiths, with engineers Bennett and Shiers, set off on their marathon flight in a Vickers Vimy. Four other crews competed for the prize, but all crashed or damaged their aircraft.

With the modest ceiling possible in such aircraft, inevitably they had to fly through their share of bad weather, and at makeshift landing-grounds there were problems with water-logged or soft surfaces. Often the fields were of worryingly marginal length, and at one a runway of bamboo matting had to be laid before departure could be considered.

They reached Darwin after 27 days and 20 hours, so qualifying for the prize, before completing their flight to Adelaide. Both men were knighted for their achievement.

In 1922 Ross was killed preparing a Vickers Viking amphibian for a round-the-world attempt. In a sad coincidence, **JOHN ALCOCK** who also rose to fame in a Vimy likewise died in a Viking.

Keith held a number of senior posts in airlines and industry. Their Vimy is preserved at Adelaide Airport.

SOPWITH, SIR THOMAS (1888-1989) – founder of the Sopwith Aviation Company. By the time Sopwith earned his flying certificate in 1910 he had already made his mark in motoring, sailing and ballooning. He soon became a leading racing pilot, earning a reputation for bravado, as in one account "when he skimmed within ten feet of the sheds it was considered he was going ahead somewhat thoughtlessly."

In December 1910 he won a prize for the longest flight from Britain to the Continent, reaching 177 miles (283 km) in his Howard Wright biplane (no link with the Wright brothers). The flight was an act of faith, for his ENV engine had never before run properly for more than 10 minutes, until long hours by his engineer Fred Sigrist persuaded the cantankerous engine to behave.

Sopwith made some of the best-known wartime British aircraft, including the Pup, the '1½ Strutter', the Triplane, and the famous Camel, so-called because of a hump to house the two guns. Tricky and unforgiving if mishandled, the Camel was still highly effective and accounted for no less than 1,294 enemy aircraft.

The coming of peace brought an icy blast of recession to aircraft builders and Sopwith closed in 1920. However it was reformed under the Hawker name but with Tom Sopwith continuing as Chairman, a post which he held until 1963.

Sopwith, Sir Thomas *(continued)*

A momentous decision of his was to order the Hurricane into production before any authorisation was received from the Air Ministry, saving vital months.
His death at the age of 101 marked the passing of the last of the early British pioneers.

Taking their own Medicine

Sopwith's first design was called the Tabloid. The makers of a pill of that name protested, but they were over-ruled and could do little except, it was suggested, "take a few of their own tabloids." It didn't harm the Tabloid either – it won the 1914 Schneider Trophy.

SPAATZ, CARL ('Tooey') (1891-1974) – wartime head of the American Eighth Air Force. In pre-war flying Spaatz had set an endurance record (now discouraged) of 150 hours 40 minutes in a Fokker trimotor, using in-flight refuelling.

In 1942 he was placed in command of the USAAF Eighth Air Force, a critical post charged with the daylight bombing raids over Europe. After D-Day he became commander of US Strategic Air Forces.

In 1945 Spaatz was promoted to general and handled the final operations against Japan. He took command of the whole Army Air Force in 1946 and was involved in the creation of a separate US Air Force in 1947, becoming the first Chief of Staff. He retired a year later.

Geoffrey Stephenson made his mark on flying history with the first glider crossing of the English Channel, on 22nd April 1939. His Kirby Gull was of low performance by today's standards.

STRINGFELLOW, JOHN (1799-1883) – 19th century experimenter. Stringfellow tested a series of models, at first with Samuel Henson. A steam-powered model of 1848 has often been credited as the first heavier-than-air (model) flight. The consensus now is that it did not really fly.

At an exhibition at the Crystal Palace in 1868, Stringfellow displayed a triplane. It never flew but it inspired others to come, notably **Chanute** and **Lilienthal.**

SUETER, SIR MURRAY (1872-1960) – instrumental in starting British naval aviation. As Director of the Air Department at the Admiralty in 1912, Sueter ordered some small airships and was largely responsible for setting up the new Royal Naval Air Service and providing it with aircraft.

The RNAS generally acquitted itself well in wartime, performing many heroic actions, but service infighting led to Sueter being replaced in 1915. In later posts he took over aircraft procurement, while on the ground he was one of the originators of the tank. Further controversy led to loss of these posts too in 1917, but his achievements were later recognised by award of a knighthood in 1934. He served as an MP from 1921 until 1945.

SUMMERS, JOSEPH ('Mutt') (1903-54) – Chief Test Pilot for Vickers and Supermarine. Only six months after joining his first squadron, in 1924, Summers was selected for RAF test flying at Martlesham Heath, a reflection of how well his piloting skill was regarded. In one incident, a Bulldog refused to recover from a spin and he stood up ready to bale out. The change in his position induced the aeroplane to recover and he promptly sat down again!

He became Chief Test Pilot of Vickers in 1928. The company were then acquiring Supermarine so his duties covered both firms. He handled 30 first flights including the Wellington (1935), Spitfire (1936), Viking (1945), Viscount (1948) and Valiant (1951).

He is best known for handling the Spitfire on its first flight, on 5th or 6th March 1936. After the 15 minute flight he requested "I don't want anything touched". By this he meant he wanted no adjustments before the next flight, not as is often supposed the quite unrealistic idea that it was already perfect. For some reason he is often referred to as "the veteran test pilot" in connection with this flight, a description surely more apt when he retired 15 years later.

TANK, KURT (1898-1983) – German aircraft designer. Originally with Rohrbach, the pioneer in modern construction methods, Tank's best-known designs were prepared for Focke-Wulf. He personally flew the advanced four-engine Fw200 Condor airliner on its first flight in 1937. He was one of the last designers of large aircraft to handle test flying as well. It showed its capability the next year with a non-stop flight from Berlin to New York in 25 hours and another to Tokyo in 48 hours with three stops, although this feat was marred when the aircraft crashed on the way back.

Less distinguished were the roles of the Condor as personal transports for Hitler and Himmler, and as 'the scourge of the convoys' in the war.

His Fw190 radial-engine fighter was one of the best of the war, and outclassed most Allied types for a time when it entered service in 1941.

After the war Tank designed jet fighters in Argentina and India. The only one built in quantity was the Indian HF-24 Marut, first flown in 1961, but it was not his most inspired design.

TAYLOR, CHARLES (c1867-1956) – built the engine for the Wright *Flyer*. A machinist who had been with the local electricity company, Taylor joined the Wrights in 1901 to mind their cycle shop while they were testing their gliders at Kitty Hawk.

Evidently his skills went beyond bicycle maintenance, for he built the engine for the *Flyer* in six weeks, machining even complex parts like the crankshaft with the limited facilities in their workshop. Whether Taylor was involved in the design is unknown.

After the historic first flight, Taylor remained with the Wrights while they tested their improved biplanes. He then lapsed into obscurity until Henry Ford had him traced while preparing the Wright museum in the 1940s. He was found working on the shop floor at North American.

Power for the Wrights

There has been much argument over the Wright Brothers' engine. Some have claimed it was an adapted Pope-Toledo car engine, while others dismiss this as nonsense. Whatever its basis, they would surely have studied car engine practice.

It is generally accepted that they turned to making their own after various manufacturers had told them they could not meet their requirements.

The engine was largely aluminium for lightness, with the four 'in-line' cylinders mounted horizontally. There were no sparking plugs, simple 'make-or-break' contacts being used.

Only one speed was possible. It gave about 12 hp initially, decaying after a few minutes. It was actually rather inefficient (Charles Manly's unit for Samuel Langley gave 52 hp for less weight), but it was enough to make history.

TAYLOR, SIR GORDON (1896-1966) – Australian trail-blazer, and flying-boat captain. Taylor flew in both world wars, in the first as a fighter pilot and in the second flying Liberators and Catalinas. In the 1930s he accompanied **Kingsford-Smith** on several of his epic flights.

In post-war years he made several long-distance flights including a double crossing of the Pacific between Australia and Chile in 1951, flying a Catalina *Frigate Bird*. He liked to time his arrivals within seconds of the schedule predicted before departure.

In 1954 he bought a Sandringham flying-boat and operated a cruise service among the South Pacific islands, surely one of the most delightful experiences in the history of air travel. It ended due to financial and bureaucratic reasons.

Oil Change, Sir?

Before a planned crossing of the Tasman Sea with Kingsford-Smith and co-pilot John Stannage, Gordon Taylor found the other two men with an engine in pieces repairing a fault. This worried him as he knew neither was qualified in engine maintenance.

During the flight, one wing engine of Southern Cross had to be shut down because of propeller damage. Soon afterwards the other wing engine started losing oil and would clearly seize before landfall. The nose engine alone could not sustain them.

In a feat of extraordinary courage, Taylor climbed onto the wing strut, drained oil from the stopped engine, then climbed out the other side to fill the ailing one. As if once was not enough, he repeated the process five times, earning a well-deserved George Cross. Fortunately the nose engine ran flawlessly – it was the one which had been in pieces the previous evening.

TEDDER, SIR ARTHUR (1890-1967) – senior British commander for the invasion of Normandy. A combat pilot from 1916 till 1918, Tedder held various squadron commands before becoming Director of Training in 1934. A spell in command in the Far East followed, then he turned to research and development in 1938. He was one of the few to support **Frank Whittle's** jet programme.

In 1941 Tedder was put in control of Middle East RAF operations. His leadership made sufficient impression for General Eisenhower to appoint him as the senior British commander for Operation Overlord, the invasion of Normandy. After the landings, he was responsible for all Allied air forces in the drive into Europe.

He was Chief of the Air Staff from 1946 to 1950.

TEMPLE, FELIX DU (1823-90) – flew models in the 1850s. Clockwork and steam powered models were flown in 1857 and 1859, with some success. Du Temple then built a full-size steam-powered aeroplane of prophetic design, even featuring a retractable undercarriage! In 1874 it made a brief hop after a downhill run. As others later found, heavy steam engines would never find their place in the sky.

TEMPLER, JAMES (1846-1924) – ran British military aviation for 20 years. Templer set up the first army ballooning units in 1884. Their first real test came with the Boer War, when they proved their value and acquitted themselves well. Strangely enough, Templer was not placed in command of the balloons for this conflict, but was put in charge of rather heavier equipment - steam traction engines.

Before long he was back with the balloons, and he also encouraged work on kites and airships. He was well respected, but in 1906 he was replaced after a probably unsubstantiated charge against a subordinate. He was still retained as an advisor and he helped see one of his favourite projects into the air, the well-designed *Nulli Secundus* airship.

TERESHKOVA, VALENTINA (1937-) – first woman in space. The flight was aboard *Vostok 6*, launched on 6th June 1963. She remained in space for just under three days.

THOMAS, GEORGE HOLT (1868-1929) – founder of Airco and Gloster. Holt Thomas became involved in aviation by offering prizes via his father's newspaper, the *Daily Graphic*, and organising air 'meets'.

In 1912 he formed the Aircraft Manufacturing Company, or Airco, initially to build Farmans. In 1913 he engaged **Geoffrey de Havilland** as designer, starting the line of famous 'DH' types.

In 1917 he was one of the founders of the Gloucestershire Aircraft Company, later shortened to Gloster, which was set up to expand wartime output of Airco designs, Bristols and Nieuports.

In 1919 Holt Thomas started airline services between Hounslow and Paris under the name Air Transport and Travel, using DH4 and DH9 'airliners'. These were among the first international services in the world. Enterprising they may have been, but profitable they were not, as soon became all too clear. The service ended in December 1920. Times were also hard for manufacturers and Airco closed with the loss of wartime business.

THOMAS, SIR MILES (1897-1980) – aircraft production manager and airline chairman. Thomas served with armoured cars in the First World War, transferring later to the RFC. He saw active service in the Middle East and held training appointments.

During the inter-war years he was a motoring journalist before joining Morris Motors. On the outbreak of war, the company's factories were turned over to aircraft production and Miles Thomas found himself charged with organising this vital output. The sites included a new plant at Castle Bromwich which became the principal source of Spitfires, building up to a peak of 350 a month, and later a major supplier of Lancasters.

A break from airline posts followed before he joined BOAC in 1948, becoming Chairman a year later. Over the next seven years he transformed the airline from a hopelessly unprofitable concern flying a mix of adapted bombers into a major international force. He found the airline grossly overstaffed, but sensitively saw each of those to be made redundant personally.

He bought more economical airliners, notably the Canadair Four, or Argonaut as BOAC called it, and started the jet age with the Comet in 1952. He had all the trauma of grounding the Comet after its accidents too. He left in 1956 to join a chemical company.

As chairman of nationalised BOAC, Sir Miles knew the ways of politicians. He wanted to order about 20 Argonauts (Canadian-built Douglas DC-4s with Merlin engines) but knew the government would cut the order back. Rather than ask for 20, he stated his needs as 22, making it seem the result of detailed calculations. It worked – he received all 22.

THOMSON, CHRISTOPHER (Lord Thomson of Cardington) (1875-1930) – promoter of the passenger airship. A soldier with experience dating back to the Boer War, Thomson became Secretary of State for Air in the first Labour government in 1924. He brought great enthusiasm to the job, supporting all forms of aviation, military and civil.

Among the projects he backed was the building of the R-100 and R-101 airships to provide empire air services. He was out of office during much of their development period, but returned in 1929. Both airships were in trouble and had to be 'stretched' to lift a worthwhile payload, among a myriad of problems.

Sir Peter Masefield has researched the story of these airships and overturned the long-held view that Thomson was inept and largely responsible for sending an unairworthy R-101 to inevitable disaster. He found that the minister acted responsibly, although attitudes to safety were very different in 1930 from those of today. It is unlikely that Thomson would have been totally reckless towards the safety of R-101 – he was, after all, travelling on it when it crashed.

TIBBETS, COL. PAUL (1916-) – dropped the Hiroshima atomic bomb. Tibbets was charged with training a unit for the special needs of the new weapon, as well as leading the mission on 6th August 1945. He flew a Boeing B-29, *Enola Gay*, named after his mother. The understandable suspense was further increased by the knowledge that the Uranium 235 device, the *Tall Boy*, had never been tested. After releasing the bomb, he turned sharply away to minimise the blast waves, but still their severity jolted the crew.

To Bomb or not to Bomb?

"My God, what have we done?", exclaimed co-pilot Robert Lewis after the explosion. The decision to use the atomic bomb followed long soul-searching. It has been debated ever since, but an invasion of Japan would have cost hundreds of thousands of lives, if not millions.

Tragic though the loss of lives was, it is probable that many times the number were saved by the bomb. Both Tibbets and Leonard Cheshire, British observer on the second raid, were always adamant that the decision was justified.

TILTMAN, HESSELL (1897-1976) – founder of Airspeed. Tiltman started in a heavier field of engineering than aviation - working on the Quebec Bridge (perhaps not the ideal apprenticeship as it collapsed twice before completion). He joined Airco, then moved to de Havilland. He acted as an observer on test flights, but an angry scene occurred when he tackled the managing director about the firm's practice of giving the pilot, but not the observer, a parachute!

A spell on the R-100 airship followed, but he was out of a job when the programme was halted after the R-101 disaster.

Tiltman and Neville Shute Norway, the author who wrote under his first two names, set up the Airspeed company. Their first design, a small airliner called the Ferry for **Alan Cobham**, flew six months later. Just over a week after its first flight it was carrying passengers.

The Courier followed, breaking new ground in Britain with its retractable undercarriage. Sceptics who told Tiltman it would not be worth the complication were confounded when the Courier flew 37 mph (59 kph) faster with the wheels raised. He enlarged it to take twin engines as the Envoy, and received the accolade of an order from the King's flight. In military form, over 8,000 were made as the Oxford trainer. A similar number were made of Tiltman's last design, the troop-carrying Horsa glider, the backbone of several airborne landings including Arnhem.

Airspeed was bought by de Havilland in 1942. Tiltman left to form a consultancy business.

Now You Need it, Now You Don't

When Tiltman submitted the Ferry for certification, the inspector told him that as a ten-seater it needed a radio. The only way to fit a bulky radio of the day was to remove a seat. When the inspector returned, he declared that as the Ferry was now a nine-seater no radio was needed!

TRENCHARD, SIR HUGH ('Boom') (1873-1956) – founder of the Royal Air Force. Trenchard had a long career as a soldier, and had sustained serious injuries in the Boer War, before starting flying. He learned with **Tom Sopwith's** school in 1912, and thereafter promotion was swift, so much so that by 1915 he was Commander of the RFC.

Trenchard worked hard for better equipment and pressed for more offensive operations. Without doubt he was energetic and in many respects effective, but he was also attacked for being callous – his policy of 'no empty chairs' to avoid harming squadron morale also meant young pilots going into action pitifully untrained, and surely he of all people could have ensured aircrew at least had the benefit of parachutes?

'Boom' Trenchard – architect of the RAF

In 1917 he became Chief of the Air Staff, a post he held with one short break until 1929. He campaigned for the creation of a separate air service, which duly came to pass with the creation of the RAF on 1st April 1918. In peacetime he had to fight for its continued existence.

In the 1920s he was able to show the effectiveness and economy of air power in containing 'brushfire' wars in the Middle East and elsewhere with a fraction of the manpower ground troops would have needed.

He founded the Auxiliary Air Force, the University Air Squadrons, and started the Hendon air displays. He saw the vital need for training and established the Staff College at Andover, the Cadet College at Cranwell, and the Apprentices School at Halton. He strongly encouraged technical research, and built as firm a foundation as possible for the RAF within tight budgets.

After retiring from the RAF, Trenchard became Commissioner of the Metropolitan Police, once again placing an emphasis on training, setting up the Police School at Hendon.

TRIPPE, JUAN (1899-1981) – founder of Pan American Airways. The airline started in 1927, flying between Florida and Havana with Fokker VII/3 trimotors. By cultivating political contacts and using some ruthlessness, by 1930 Trippe had extended Pan American services into much of South America. In an astute publicity move, he employed **Charles Lindbergh** as technical consultant and to handle route-proving flights.

Services were expanded with longer-range Sikorski S-42 and Martin M-130 flying-boats. A Pacific service started in 1935, and in 1939 he launched a transatlantic route with new Boeing 314 flying-boats. This service continued during the war.

In 1946 he changed to landplane transatlantic services, with Constellations, adding the double-deck Boeing Stratocruiser in 1949. By the early 1950s he saw the jet age coming and ordered three Comet 3s, later cancelling them after the Comet 1 grounding. Now the two American jet airliners, the Boeing 707 and Douglas DC-8 became candidates for his orders. His decision was awaited anxiously, for whichever he chose, it was likely the other would fall by the wayside. In the end the wily Trippe confounded everyone by ordering both!

In the 1960s he prepared Pan American for the wide-body age, but by the time the Boeing 747 entered service in 1970 he had retired.

Now, in the 1990s the unthinkable has happened – the great Pan American, once a symbol of its nation's might, is no more.

TRUBSHAW, BRIAN (1924-) – first British Concorde test pilot. Trubshaw joined Vickers as a test pilot in 1950, having served in the RAF since 1942. In 1964 he became Chief Test Pilot of the commercial division of the British Aircraft Corporation.

He flew the first British-assembled Concorde on 9th April 1969. He handled much of the flight testing over the next seven years – Concorde was the most thoroughly tested airliner in history, six of the 20 built being used exclusively for testing.

A series of supersonic Concorde flights were made over land to assess the effect of the 'boom' on the ground. Times were announced in advance. At the advertised time of the first test, complaints came in on cue. Alas, Concorde was still on the ground, delayed by a fault.

The Americans, too, had problems with supersonic trials, in their case over Oklahoma. An unforeseen reaction to the boom came from skunks, which responded in the only way they knew.

TSIOLKOVSKY, KONSTANTIN (1857-1935) – visionary of space travel. This remarkable, largely self-taught and partly deaf Russian schoolteacher wrote scientific papers on aeronautics and spaceflight before 1900. By 1903 he had proposed liquid fuelled and even multi-stage rockets for manned spaceflight, truly visionary ideas at a time when the first aeroplane had yet to fly, let alone spacecraft. Like many of prophetic ideas, his work was not properly recognised until late in his life.

TUCK, ROBERT STANFORD (1916-87) – leading RAF fighter pilot and tactician. A pre-war pilot, Stanford Tuck survived a collision in 1938 which killed the other pilot. During the war he shot down 29 enemy aircraft, his piloting skill being matched by his outstanding shooting accuracy.

He was one of the leading advocates of more flexible tactics and formations. The RAF had tended to cling to the tight formations so impressive at prewar airshows, whereas German pilots had learned in Spain the value of looser units.

In September 1940 he took command of a squadron which had taken heavy losses but quickly restored its efficiency. He survived baling out and a ditching, but early in 1942 was shot down over enemy territory and spent the rest of the war in captivity.

TUPOLEV, ANDREI (1888-1972) – Russian aircraft designer. The first Russian all-metal aircraft was Tupolev's ANT-2 of 1924, which owed much to Junkers designs of the day.

In the inter-war years Tupolev was best known for a line of outsize aeroplanes, culminating in his ANT-20 *Maxim Gorki*, an eight-engine monster of 206 ft (63 m) span. This product of Stalinist Russia served no more honourable purpose than disseminating propaganda, and illuminated signs and loudspeakers were fitted to broadcast slogans. It flew in 1934, but met with disaster a year later when a fighter doing unauthorised aerobatics collided with the *Maxim Gorki*, killing all 47 on board.

Tupolev was in trouble too, for he was arrested in 1936 for allegedly passing data to Germany, but in a bizarre arrangement he was allowed to continue design work 'inside'. His resulting Tu-2 bomber was promising enough to secure his release in 1943.

In post-war years Tupolev moved into the jet age with his Tu-16 'Badger' swept-wing bomber, which entered service in 1955. An airliner development, the Tu-104, became the world's second commercial jet in 1956, and the Tupolev company has been a major Russian jet airliner builder ever since.

Andrei Tupolev handed over to his son, Aleksei. Before he died, he saw Russia's supersonic airliner, the Tu-144, fly on the last day of 1968, but it is perhaps as well he did not see its later problems and withdrawal from brief service.

TURCAT, ANDRÉ (1921-) – handled first flight of Concorde. One of France's leading test pilots, Turcat had made his mark setting a series of closed-circuit records in a Nord Griffon, a research aircraft with a mixture of turbojet and ramjet power.

He made the maiden flight of Concorde on 2nd March 1969 and thereafter was in charge of the French end of flight testing.

TWISS, PETER (1921-) – Chief Test Pilot of Fairey and holder of world speed record. Twiss was a wartime naval pilot, whose service included manning Hurricanes aboard merchant ships for convoy protection. He also had a spell on night-fighters.

At Faireys he flew the Firefly, Gannet and the two Fairey Deltas. The first of these, the Fairey Delta One, was a tricky beast to fly and unpopular with pilots, but its successor was an excellent research aircraft. On 10th March 1956 Twiss flew the Fairey Delta Two to a new world speed record of 1,132 mph (1820 kph), some 310 mph (495 kph) faster than the previous record.

Twiss retired from test flying in 1960 and transferred to the company's marine division.

The One-Mission Fighters

Twiss was one of the courageous pilots who served on merchant ships, ready to be catapult-launched in a Hurricane to defend the convoy. The principal quarry was the Focke-Wulf Condor which shadowed convoys reporting their position.

The catch was that the Hurricane had to ditch afterwards, and rescue would have been uncertain if the ships were under threat. Thankfully, only a few missions were flown, but there was a deterrent value.

UDET, ERNST (1896-1941) – fighter pilot, aerobatic artiste, and Luftwaffe technical officer. Udet started combat flying in 1916 and became the second-highest scoring German pilot with 62 victories. In June 1918 he was one of the first to use a parachute in emergency.

After the war he designed and built light aircraft, but his real métier was as an aerobatic and stunt pilot. Specialities included picking up objects from the ground with a wing-tip and flying through open hangars (it is unlikely any airshow organiser would permit such antics today). He flew such dramatic stunt sequences for films that they were widely believed to have involved trick photography.

For some time he resisted offers of senior military posts from the Nazi hierarchy, but in 1936 he succumbed and became Inspector of Fighters. He was involved in the selection of aircraft types and it was largely his decision which put the Messerschmitt Bf109 into production, despite initial coolness towards it. He was also an enthusiast for dive-bombers and promoted the Junkers Ju 87 Stuka.

He was unhappy 'flying a desk' and never really settled into ground appointments or working with Nazi leaders. The one talent he displayed on the ground was sketching, and his drawings of personalities and aircraft are now collectors' pieces. He clashed repeatedly with **Erhard Milch** and finally shot himself in November 1941.

UWINS, CYRIL (1896-1972) – Chief Test Pilot, Bristols. In a remarkable test flying career, Uwins made over 50 first flights between 1919 and 1947.

Notable maiden flights included those of the Bulldog (1927), Bristol 142 *Britain First* (1935), Blenheim (1936), Beaufort (1938), Beaufighter (1939) and Bristol 170 Freighter (1945).

In 1920 he flew the Bristol Bullet, the first aircraft powered by a Bristol engine, to second place in the Aerial Derby, giving the new line of engines a welcome boost.

He gave the engine division another coup by setting a world height record of 43,976 ft (13,404m) on 16th September 1932. He was flying a Vickers Vespa with a virtually standard Pegasus engine.

VOISIN, GABRIEL (1880-1973) and **CHARLES** (1888-1912) – pioneer French aircraft builders. A lecture by Octave Chanute to the Aero Club de France in 1903 on the Wright Brothers' gliding trials stimulated interest in heavier-than-air flight. A prominent member, Ernest Archdeacon, commissioned Gabriel to build a glider similar to those flown by the Wrights, as far as possible given that the American brothers guarded their secrets closely. At the time Gabriel was an architecture student who had made a name for himself building kites.

His brother Charles joined him to form the company Voisin Frères. Other gliders followed, some mounted on pontoons for towing from water. On one launch Gabriel was nearly drowned. They sold the gliders to other pioneer aviators, including **Santos-Dumont** and Blériot, with whom they formed a partnership for a time.

The brothers advanced to powered machines of 'boxkite' layout in 1907. They were among the first to offer aeroplanes for sale. The **Farmans** were early clients, but fell out with the Voisins when their order was diverted to **Moore-Brabazon**, who was in no way involved.

Voisin designs were 'inherently stable', so if a bank developed they would right themselves. There was no lateral control, so change of direction was by skidding turns on rudder alone. Not until 1910 did the brothers fit ailerons.

Charles was killed in a car accident in 1912. Voisin left aviation after the war. In later life he made extravagant claims that it was he, rather than the Wrights, who had invented the aeroplane.

VOLKERT, GEORGE (1891-1978) – designer for Handley Page. Volkert's long career ranged from 100 mph (160 kph) biplane bombers to early studies for the 600 mph (1,000 kph) Victor.

He joined the company in 1912, having previously known **Frederick Handley Page** at Northampton College. He handled design work on the O/100, O/400 and V/1500 bombers.

He left the company for a year in 1922 to serve on a British aviation mission to Japan.

Back with 'HP', he designed a series of biplane airliners, the best known of which was the HP42, one of the first four-engine airliners and almost a symbol of prewar Imperial Airways.

Volkert again left for a spell between 1931 and 1935. His place was taken by German-born Gustav Lachmann, who designed the Hampden bomber but was interned for a time during the war because of his origins. He was later released and rejoined the company.

Once more returned to the fold, Volkert produced his most famous creation, the Halifax heavy bomber. Although somewhat eclipsed by the Lancaster, the 'Hallibag' formed a most important element of the bombing effort and was also used for anti-submarine work and glider towing.

Nominally Volkert retired in 1945 but he continued as a technical consultant.

VOUGHT, CHANCE (1890-1930) – American aircraft builder. After learning to fly in 1910, Vought had the audacity to modify a Wright biplane by replacing the pusher propellers with 'tractor' units. He confounded those he thought a Wright design could not be improved by achieving far better performance.

He started his own company with a relative, Birdseye (yes, really!) Lewis in 1917. A range of fighters and trainers emerged.

As Chance Vought Corporation, the company went on to build one of the best naval fighters of the Second World War, the F-4U Corsair. 12,571 were made, serving both the US and Royal Navies. Vought himself did not live to see his company's finest contribution to history, for he died at the early age of 40, but he had laid the foundations.

WALLIS, SIR BARNES (1887-1979) – aircraft, airship, and weapons designer, and pioneer of 'swing-wings'. Barnes Wallis began his career in marine engineering. After a period of wartime army and RNAS service he became Chief Designer for the airship division of Vickers, and later for its subsidiary the Airship Guarantee Company. He designed the large rigid airships R-80 and R-100. The latter made a successful double Atlantic crossing in 1930.

Problems were lurking on the R-100. After the R-101 disaster she was scrapped and Barnes Wallis was out of a job. He moved to the aircraft side of Vickers, and adapted airship construction methods as his 'Geodetic' structures. First used on the Wellesley bomber, the technique came to fame on the Wellington, produced in greater numbers than any other British bomber. Geodetic structures were strong and withstood heavy damage, but the method disappeared with the arrival of pressurisation.

He conceived the bombs released on the famous dams raid on 16th May 1943. The cylindrical bombs bounced over the water until they reached the dam, where they exploded after sinking well below surface level. He was deeply upset by the casualties on the operation.

Still with bombs, he developed the *Tallboy* and *Grand Slam* bombs, which fell at great speed and had toughened casings to penetrate targets below ground or under concrete. The ten ton *Grand Slam* was the heaviest bomb of the war.

In 1945 he began work on variable geometry, or swing-wing, aircraft. His objective was a long-range airliner. He conducted years of research on projects loosely named *Wild Goose* and *Swallow*. His ideas were used in the American F-111, but it was not until 1974 that a European aircraft, the Tornado, flew with his principles.

Stretching a Point

Barnes Wallis' geodetic construction proved sturdy and easy to make, but it did not take kindly to glider-towing. After one trial the Wellington was said to have become 1½ ft (½ m) longer!

WARSITZ, ERICH – flew first jet aircraft. The historic first flight of the Heinkel 178 took place on 27th August 1939.
He also flew Heinkel rocket aircraft, including the He 176 in 1939 and the He 112, in which he was fortunate to have survived when it exploded.

WENHAM, FRANCIS (1812-1908) – built first wind tunnel. In 1866, at the first meeting of the Aeronautical Society, Wenham presented a paper *On Aerial Locomotion.* He outlined his theoretical work on the lifting characteristics of aerofoils and described gliding trails he had carried out in 1858-9. He had conducted some of his trials after dark – ridicule was a worry to the pioneers. He escaped once when he was blown over in his glider.

He was encouraged to expand his work and build his wind tunnel in 1871 with John Browning, another Society member. Airflow was provided by a steam-driven fan.

Two major findings emerged from his work, the advantage of high aspect ratio wings as opposed to the almost square ones of other experimenters, and the structural advantages of multiple wings rather than monoplanes. These conclusions were to prove important, for **Octave Chanute** made them widely known and led to the successful flights of the Wright brothers.

WHITCOMB, RICHARD (1921-) – aerodynamicist. In the 1950s certain supposedly supersonic American military aircraft were failing to achieve anywhere near their design performance. The worst was the Convair F-102 which was totally unable to reach Mach 1 and worst of all, was already in production. NACA aerodynamicist Whitcomb proposed his 'Area Rule', in which ideally the cross-section of a supersonic aircraft should remain as near constant as possible. To compensate for the sudden increase in section where the wings join, bulges could be added to the aft fuselage. Applied to the F-102 it worked spectacularly and saved the day for that programme.

Later he developed the 'supercritical wing', looking almost like an ordinary aerofoil upside-down, which yielded higher efficiencies at high subsonic speeds, and the 'winglets' now widely used to reduce the energy loss in wing-tip vortices.

WHITE, SIR GEORGE (1854-1916) – founder of the Bristol Aeroplane Company. A Bristol businessman who started the first electric tramway network in Britain, White acquired a number of railway and tram companies throughout Britain.
After seeing flying demonstrations in France, he set up the British and Colonial Aeroplane Company, later Bristol, in 1910. His first success was the Bristol Boxkite, a copy of a Farman. At the army exercises in 1910 he had two of them demonstrated, one with radio, to show the use of aerial reconnaissance.

Astutely, he provided flying training for purchasers, and by the outbreak of war had sold some 200 aeroplanes, making his company one of the largest in Britain.

In the year he died, the most important wartime Bristol design flew, the F2A Fighter. It was the beginning of a long line of fine Bristol designs. He was succeeded by his brother, Samuel White, and for most of the history of the company the chairman was a member of the family.

Sir George reputedly sited his airfield at Filton because it was at the end of a tram route and would provide maximum revenue for the trams from employees and spectators.

WHITTLE, SIR FRANK (1907-) – jet engine pioneer. As the subject for his thesis at the RAF College, Cranwell, in 1928, Whittle proposed the use of gas turbines for aircraft propulsion. He did not invent the gas turbine – industrial units already existed – and others were working on aircraft applications, but the radical difference with Whittle's schemes was that they used a pure jet. The others proposed turbines driving propellers (turboprops).

He took out his historic jet patent on 16th January 1930. In the early 1930s his RAF postings limited the work he could do, and he actually let his patent lapse in 1935 for want of the £5 renewal fee, but in 1936 he formed Power Jets to turn his ideas into hardware. He opted for the simple centrifugal flow layout rather than the more complex axial type.

On 12th April 1937 Whittle ran his first engine. It quickly ran out of control and others nearby fled, but he stayed to shut off the fuel – a pool of fuel had collected in the combustion chamber. Later runs were successful, subject to inevitable problems in running an entirely new prime mover. His own position, as a serving RAF officer while running a business, was a strange one and caused its own share of problems. Progress was a constant battle against official indifference and lack of funds.

The first flight was in the Gloster E28/39 on 15th May 1941, in the hands of test pilot Gerry Sayer. It seems extraordinary, even allowing for wartime secrecy, that no official film or photographs were taken of the historic event. It was not until after the war that it was realised that the Heinkel 178 jet had flown in 1939.

Even after the jet had proved itself in the air, progress was slow due to limited support and the division of work. Air Ministry policy decreed that Power Jets built development engines only, the main production effort being allocated to the Rover company.

Relationships between Whittle and Rover soured, partly due to the latter's insistence on making their own changes to design. Matters improved when Rolls-Royce took over from Rover in 1942, and at last the first production engine, now named Welland, started to appear for the Meteor. Power Jets was later absorbed into the Royal Aircraft Establishment.

In 1948 Whittle resigned from the RAF, and in the same year was knighted. He took post-war consultancies with BOAC, Bristol Siddeley and Rolls-Royce.

In the 1950s he devised a turbo-drill for deep bores, to avoid the sometimes miles of shafts normally used. This promising idea ran into technical and commercial problems and was abandoned after 13 years work in 1966.

Sir Frank emigrated to the USA, somewhat disillusioned with the lack of backing he had so often met in his career.

WILLS, PHILIP (1907-78) – the leading name in British gliding for several decades. A light aircraft pilot in the 1920s, Wills was introduced to gliding in 1932 and thenceforth devoted himself to the sport. He was British National Champion no less than five times, and World Champion in 1952.

On the ground, he worked energetically for the movement, becoming Chairman of the British Gliding Association in 1949. He was also an accomplished writer whose books relate the challenges and above all the humour of international competition.

Neil Williams, one of the greatest of British aerobatic pilots, performed an almost unbelievable feat of flying and one which must almost certainly be unique.

Whilst practising aerobatics in a Zlin, one wing failed and started folding upwards. Instantly he rolled inverted to relieve the load on the wing, returned to the airfield upside down, finally rolling upright within a few feet of the ground. (Try working out which direction of roll would ease the strain on the wing, remembering he had to decide instantly).

Ironically he was killed on what should have been a routine ferry flight of a CASA 2111 (Spanish-built Heinkel 111) when he struck a mountain in the Pyrenees.

WRIGHT, WILBUR (1867-1912) and **ORVILLE** (1871-1948) – first sustained, controlled, heavier-than-air powered flights. The Wright brothers succeeded where others failed because they recognised the need to control an aeroplane, they realised they had to learn the art of piloting as well as building a machine, and they carefully studied wing and propeller design. At the time they took an interest in flight they were running a cycle shop in Dayton.

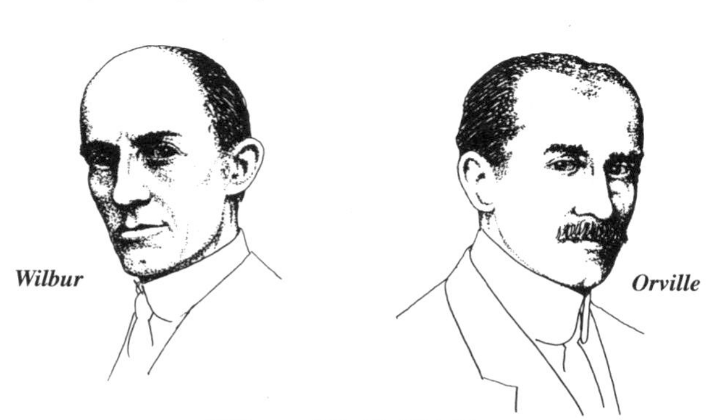

Wilbur and Orville Wright –
first to fly a successful aeroplane

It was Wilbur who first became interested in flying after reading about the work of **Lilienthal** and **Chanute**. He quickly grasped the need for good control and that shifting body weight would not be enough. By watching birds, he deduced that they must be using some form of wing-tip flexion. It is believed he also studied a book with photographs of raptors which confirmed his deduction.

Form this conclusion, and a chance observation of the way an empty cycle-tube box could be twisted, he devised his system of 'wing-warping'. In principle it works in the same way as the now universally used ailerons – the difference is purely mechanical. Here was the first Wright breakthrough; some designers were still not providing lateral control ten years later.

The brothers started with a biplane glider, which drew on **Chanute's** work on structures, in 1900 at Kitty Hawk, North Carolina. The spot was chosen for its plentiful winds, low hills, and seclusion. The glider was tested first as a kite, with Wilbur lying prone on the lower wing. The wing-warping proved excellent but the lift from the wings was disappointing.

The following year they tried a larger glider and Wilbur started gentle turns. Loss of control occurred in the turns which puzzled them until they reasoned that there was more drag on the raised wing.

These trials made them suspect the data published by Lilienthal and others. They built a simple wind-tunnel and ran thousands of tests on aerofoils to amass their own figures.

Using these data they built a third glider in 1902, this time with a fixed vertical tail. Turns still tended to end in loss of control. Then came another great inspiration, of making the tail movable. They had conceived the method used ever since for turning an aeroplane, with co-ordinated bank and rudder. A range of successful glides followed, showing they had grasped the technique of piloting. Orville joined in the flying this season.

In 1903 they advanced to powered flight with their first Flyer. The engine was built by their shop engineer, **Charles Taylor**, and is described under his entry. The propeller design proved complex. Wilbur tried to research marine propeller theory, only to find none existed! He returned to the wind-tunnel and produced a remarkably efficient propeller, far in advance of the crude paddles used by others.

An attempted flight on 14th December ended after a few seconds. Wilbur was unable to master the sensitive elevator in time.

The day of destiny came on 17th December 1903. There was a bitterly cold 27 mph (43 kph) wind, rather higher than would normally be tolerated for a first solo today! The brothers had set up a camera on a tripod and briefed an officer from a nearby life-saving station, John Daniels, to take what would prove the most famous photograph in aviation history. It was Orville's turn, and his 12 seconds and 120 ft (37 m) is generally considered the historic first flight. Later Wilbur managed an impressive 59 seconds and 852 ft (262 m). Now, incontestably, the aeroplane had flown. The triumph was marred when the *Flyer* was blown over and wrecked.

In 1904 and 1905 trials continued with improved *Flyers,* with circuits being achieved. These flights were conducted in obsessive secrecy, and herein lay the source of problems to come.

From 1905 onwards the Wrights tried to sell their *Flyers* to the American, British and French armies, but they refused to allow demonstrations or photography until contracts were signed. Reasonably enough, the prospective purchasers wanted to see the product before buying and impasse developed. While negotiations dragged on, the Wrights remained earthbound. Their lead was slipping.

Belatedly in 1908 they laid on demonstrations, Orville to the US Army and Wilbur in France. The latter created a sensation with his display of control in the air, and licences were signed to build Wright biplanes in France, Britain and Germany. Orville's trials were successful but were marred by an accident which killed his passenger, Thomas Selfridge.

Wilbur's last public display was a dramatic flight over New York harbour in September 1909, coinciding with a naval review. From then on, they allowed their commanding lead to be whittled away, and by 1912 when Wilbur died of typhoid, the *Flyers* were looking dated. This astonishing situation resulted from their obsessive litigation over their patents, causing them to neglect improvements to the *Flyers*. It was a sad twilight to their remarkable pioneering efforts.

After Wilbur died, Orville sold his holdings in the Wright company and played little further part in aviation.

The Wright Way to Fly

The original Wright Flyer was a biplane with forward elevators controlled by the pilot's hands. Lateral control was by wing-warping worked by moving the hips in a cradle. The rudder was linked to the wing-warping control.

The engine of about 12 hp drove two pusher propellers via chains, one of which was twisted so they turned in opposite directions. Take-off was from a trolley mounted on a wooden rail set into wind. There was no catapult assistance on the first Flyer, although there was later to reduce dependence on winds.

On the early Flyers the pilot lay prone on the lower wing. Later they changed to a seated position. (In the 1950s the prone position was tried again, in an adapted Meteor. Pilots disliked it, one commenting "There is only one thing a pilot can do in that position and it has nothing to do with flying.")

YAKOVLEV, ALEXANDER (1906-89) – Russian aircraft designer. Yakovlev started with gliders in 1924, graduating to light aircraft in 1927. Until 1940 he concentrated on trainers and light aircraft, but during the war his fighters were built in vast numbers, reportedly over 36,000.

He achieved the distinction of putting the first Russian jet into service, the straight-wing Yak-15, but it was soon eclipsed by the more advanced MiG-15. In 1966 he branched out into airliners with the Yak-40, which came close to being built in the West as well.

As seems to be traditional in Russian aviation, his son Sergei followed him as designer.

YEAGER, CHARLES ('Chuck') (1923-) – first to fly faster than sound. Yeager was a wartime pilot with a score of $13^1/_2$ victories, including five in one day. He was shot down and wounded over France, escaping to Spain with the aid of the Resistance.

'Chuck' Yeager – first faster than sound, but safer than horses

On his return to America he began test flying, handling the early American jets at the Muroc dry lake test site, later Edwards Air Force Base.

When the Bell X-1 rocket research aircraft (then designated XS-1) was announced, volunteers were called upon for its testing which was, rightly, expected to be unusually dangerous. Flying started with glides from a B-29 'mother ship', building up to powered flights with progressively higher thrust.

Yeager was poised to handle the attempt to exceed Mach 1 for the first time when he was injured, not by the obvious hazards of his job but by a fall from a Mach 0.03 horse. He kept the injury secret to avoid losing his chance to another pilot. The historic flight took place on 14th October 1947. He named the X-1 *Glamorous Glennis* after his wife. Buffeting and loss of elevator control occurred as he neared Mach 1, then it suddenly became smooth as he accelerated past the magic number.

He completed 40 flight in the X-1 and improved X-1A. In the latter he reached Mach 2.4 but lost control and was lucky to survive.

Yeager returned to squadron flying and flew operationally in Vietnam. He retired in 1975.

YEAGER, JEANA – her round-the-world flight in *Voyager* is described under Dick Rutan. She had played a major role in planning the flight and building the aircraft. She had previously applied to be an astronaut under a still-born private space programme.
YOUNG, JOHN (1930-) – astronaut and first commander of the Space Shuttle. Young flew on two *Gemini* earth orbital missions and two *Apollo* moon-flights.
He flew the first Shuttle mission on 12th April 1981 with Robert Crippen in *Columbia* after many delays due to loose thermal tiles. They completed 37 orbits.

ZEPPELIN, COUNT FERDINAND VON (1838-1917) – pioneer of rigid airships. As an army officer, von Zeppelin became interested in airships for their military value. His first such craft, LZ-1 (Luftschiff Zeppelin 1) flew, rather erratically, from Lake Constance in July 1900. Like all Zeppelins, it was a rigid airship, in which the gasbags were held in a metal framework.
A series of improved airships followed, but progress was far from smooth and most were lost in accidents. It is remarkable that none of the mishaps involved loss of life until 1913.
In 1909 he started an airline service, DELAG. This first airline in the world carried around 34,000 passengers (a range of figures are quoted by different sources). No passenger was killed on these services, rather flukishly as several airships were lost in accidents.
His military airships were used for reconnaissance until 1915, when the Kaiser ordered bombing raids on cities. At first they flew with impunity above the reach of guns or aeroplanes, but not for long. On 7th June 1915 Fl. Lt. Warneford flew a Morane aircraft above LZ-37 and bombed it, a feat which won him a VC. Casualties among the Zeppelins rose and raids dwindled to a spasmodic level. Despite the hopes of their creator, they had been of little military value.
So closely was the count's name associated with airships that any large rigids tended to be called 'Zeppelins'. Much less well known were his large wartime bombers, made under the name Zeppelin Staaken after site of the factory. He retired in 1915 at the age of 77.
ZURAKOWSKI, JAN – test pilot. A Polish escapee, Zurakowski's flying was outstanding and he was selected for test flying, working successively for Martin-Baker, Gloster and Avro Canada. He originated two new display manoeuvres.
On the Meteor he devised his cartwheel, in which he climbed vertically then cut one engine to idling while applying full power to the other. The aircraft then rotated like a wheel.
In the Avro Canada CF100 fighter he again created a sensation, this time with his falling leaf descent, well described by its name.

Glossary and Abbreviations

The use of technical terms, jargon and abbreviations has been kept to a minimum, but a few are unavoidable and are listed below.

aileron: small moveable surface at each wing-tip for lateral control. Thus to bank the wings to the left, the aileron on the left wing is raised while that on the opposite wing is lowered.

airship, non-rigid: one in which the envelope shape is held by gas pressure alone.

airship, rigid: one in which the envelope is supported by a metal or wooden framework.

The term ***dirigible,*** *a now rather archaic word for an airship, sometimes causes confusion. It simply means 'steerable', from its French origins, as opposed to a balloon. It has nothing to do with being rigid or non-rigid.*

autogyro: a rotary-wing aircraft in which the rotors are turned by forward movement through the air. There is no drive from the engine to the rotors in flight, although some could engage a drive for take-off only. The spelling autogiro was a proprietary name used by Cierva.

axial flow: a turbine engine layout in which the airflow proceeds in an essentially straight line through the compressor and turbine. It is the usual type today except in very small engines.

BEA: **B**ritish **E**uropean **A**irways, the airline which handled domestic and continental flights from 1946 until 1972, when it merged with BOAC to form British Airways.

BOAC: **B**ritish **O**verseas **A**irways **C**orporation, the airline formed in 1939 when Imperial Airways and British Airways merged. The latter name was revived for its successor in 1972.

centrifugal flow: a turbine layout in which the compressor throws the air radially outwards. Most early British, but not German, jets were of this type.

prototype: an aircraft built for flight testing and development. There may be one or several, or sometimes none at all, in which case early production aircraft are used for flight testing.

radial engine: an air-cooled piston engine in which the cylinders are arranged like the spokes of a wheel. A variation was the rotary, in which the entire engine rotated to assist cooling.

RAF: **R**oyal **A**ir **F**orce, formed as a separate service on 1st April 1918.

RFC: **R**oyal **F**lying **C**orps, the air branch of the army, superseded by the RAF in 1918.

RNAS: **R**oyal **N**aval **A**ir **S**ervice, the naval equivalent of the RFC.

turbojet: the simple jet engine in which all the air passes through the combustion chambers.

turbofan: a jet engine in which part of the airflow is ducted around the combustion chamber and turbine. The term by-pass engine was used at one time for a similar arrangement.

turboprop: a turbine engine driving a propeller.

USAF: **U**nited **S**tates **A**ir **F**orce. Until 1947 it was USAAF, with an extra A for Army.